Coping Mechanisms for Climate Change in Peri-Urban Areas

S. Manasi • K. V. Raju

Coping Mechanisms for Climate Change in Peri-Urban Areas

Springer

S. Manasi
Centre for Research in Urban Affairs
Institute for Social and Economic Change
Bengalaru, Karnataka, India

K. V. Raju
Policy and Impact, Asia Program
International Crops Research Institute
for the Semi Arid Tropics (ICRISAT)
Hyderabad, Telangana, India

ISBN 978-3-030-18516-9 ISBN 978-3-030-18517-6 (eBook)
https://doi.org/10.1007/978-3-030-18517-6

This Springer imprint is published by the registered company Springer Nature Switzerland AG
The registered company address is: Gewerbestrasse 11, 6330 Cham, Switzerland

Foreword

This book is a detailed study by Dr. S. Manasi and Prof. K.V. Raju on land use change, on public-private people partnership on rural health programmes and on the environment with a special focus on understanding the role of culture in the environment, waste management and use of AYUSH systems of medicine. The authors have covered the aspects of climate changes, groundwater depletion and agricultural patterns and expanse of other resources. The focus is on traditional knowledge of ecology, protecting the environment indicating linkages between cultural practices and ecological protection, benefits of the usage of medicinal plants and the complete ecosystem which is the key for health promotion implicating climate change effects.

I believe that there is a huge difference between the healthcare service available in the city, which is world-class, and the healthcare service in the peri-urban area, which is of a very low level. Even their income levels are much better than the rural areas. Healthcare services are not in good state. The holistic approach in healthcare considers environmental, ecological, sociological and nutritional health and well-being. The focus is more on the prevention and early intervention with the treatments of homoeopathy, Ayurveda and naturopathy rather than depending only on western medical intervention. Several diseases can be prevented by nutritional support, environmental changes, proper hygiene and lifestyle modifications.

Environmental pollution and excessive usage of chemicals, pesticides and fertilisers have caused a huge impact on health, thereby causing several disease conditions including cancer. Environment and health is an important subject, which is now getting a lot of attention, and studies are showing some insights into this aspect. The traditional eating habits with locally grown foods are replaced by factory-made chemical-filled processed foods which is causing huge nutritional and health effect on the growing population. The focus on locally grown chemically free, fresh and nutritional vegetables, greens and millets is very important to improve health.

SOUKYA Foundation's experiment goal with Jadigenahalli Panchayat, Hoskote Taluk (Bangalore Rural), in establishing Dr. Mathai Rural Holistic Health Centre is to provide AYUSH (Ayurveda, Yoga and Naturopathy, Unani, Siddha and Homoeopathy) systems of medicine as an alternative to the western medicine with an aim to promote holistic health. Our experiment showed us a huge positive impact in and around several villages of Jadigenahalli Grama Panchayat. This was addressed by improving their nutrition, changing their lifestyle and introducing yoga which in turn reduced the dependence on western allopathic medicines, and these interventions are well documented in this book. This model has proved that the holistic health approach, which is more health oriented than disease oriented, will help people to take responsibility of their health with improved nutrition, proper sanitation and appropriate lifestyle habits.

I feel this study and research findings will be of help to the government in changing policies in the peri-urban areas. Besides, it will help researchers and students working on peri-urban areas, climate change and environment.

I. Mathai
Chairman, Managing and Medical Director
Soukya International Holistic Health Centre
Bengaluru, India

Preface

Peri-urban areas, as part of growing urbanisation, have drawn greater attention from planners, researchers and scientists in recent decades. Peri-urban areas comprise the characteristics of both urban and rural areas and are located in the midst of rural landscape, with emerging challenges. The ecology of peri-urban areas gets neglected in the planning process resulting in increased risks of climate change. These areas have attracted several studies focusing on peri-urban interface, investigating urban forms, spatial patterns, implications, etc.; however, there are few studies that have emphasised the role of ecological ethics and cultural practices in the context of development versus conservation conundrum. Given this backdrop, a global city like Bengaluru in India has been witnessing a wide range of changes. The local residents (including the authors) have experienced the city expansion and its magnitude and have for sure not only felt overwhelmed but also petrified of the several challenges, both for the present and the future. Our interest in this dimension was increased with several readings by urban planners, environmentalists, geographers, economists, sociologists and research studies that we were closely involved over the years.

Like other peri-urban regions, Bengaluru City also has undergone changes that may be similar to other urban areas. Climate change dimension of peri-urban areas is linked to local ecosystems of these regions. Patterns of urban expansion and growth differ extensively by region, with diverse implications for its sustainability. Given the increasing concentration of people and the extent of these areas, together with climate change projections, urban and peri-urban sustainability and safety are a growing concern. Keeping all these features in perspective, we focused on micro level dynamics, vulnerabilities and coping mechanisms of peri-urban regions. Since the peri-urban regions of Bengaluru fall under the semi-arid zone, we chose Jadigenahalli Village Council (*Grama Panchayat),* which could broadly represent a comparable situation in similar regions for an in-depth exploration as part of understanding and analysing some of the pertinent questions related to this peri-urban region.

Our book tracks down the recent changes and its influences on environment in the peri-urban contexts. It has provided insights into the complex issues of urbanisation and the associated externalities and brought to the fore the challenges faced by the peri-urban areas in terms of health, environment and climate. The book has lucidly analysed the following: (a) land use changes reflecting the impact of urbanisation on water resources, (b) intervention of a healthcare model (public-private partnership) through the Ministry of AYUSH for introducing holistic healthcare and (c) traditional healthcare laced with local cultural practices, which play an important role in local environmental protection. This empirical study ascertains that culture plays a significant role in protecting the environment and thereby in combating climate change implications.

Bengalaru, India S. Manasi
Hyderabad, India K. V. Raju

Acknowledgements

This book is largely based on multi-year research study carried out by the Center for Ecological Economics and Natural Resources and Center for Research in Urban Affairs of the Institute for Social and Economic Change, Bangalore, India. At the outset, we express our grateful thanks to our former in-charge Director Prof. MR Narayana, Prof. KS James and current Director Prof. MG Chandrakanth for their support and encouragement throughout the course of the study.

We are extremely thankful to the Soukya Foundation and its senior team, Dr. Isaac Mathai, Director, Dr. Suja Isaac, Dr. Sudha, Dr. Satheesh, Dr. Manjunath and Dr. Ganesh, all from Bengaluru, for providing us all the support, facilitating extensive discussions and providing us access to their secondary data and related documents for the study. Our sincere thanks to ASHA health workers, Gram Panchayat President and Members and respondents in Jadigenahalli who have been cooperative and responsive in providing us their views and sharing their concerns and spending enormous time in responding to the questionnaires.

We thank BR Hemalatha and KP Rashmi, Senior Research Assistants, who helped us throughout the study and made valuable contributions.

We deeply mourn the demise of Ms. S Poornima, Research Assistant, in October 2015 and gratefully acknowledge her valuable contribution to the study. She deserves enormous appreciation for all the hard work, diligence and sincerity in contributing to this study right from its inception.

Our special thanks and appreciation to manuscript reviewers and for their critical suggestions and inputs.

We thank ISEC's administration, Prof. Manohar Yadav, the then in-charge Registrar, and the then Accounts Officer Smt. Sharada, Smt. Jyothi and Ms. Niveditha for their support.

S. Manasi
K. V. Raju

Contents

1 Introduction 1
1.1 Why This Book? 1
1.2 Peri-Urban Areas 4
1.3 Evolution of Peri-Urban Areas Around Indian Cities. 6
1.3.1 Peri-Urban Ecosystems 8
1.3.2 Vulnerability in Peri-Urban Settlements 11
1.3.3 Water Crisis and Insecurity in Peri-Urban Contexts 12
1.3.4 Peri-Urban Changes and Concerns Over Food and Nutrition 13
1.3.5 Threats to Peri-Urban Forestry. 14
1.4 Urbanization in Peri-Urban Areas 14
1.4.1 Urbanisation – Influence of Development Process 14
Annexures 20
Annexure 1.1: Jadigenahalli – Peri-Urban Fringe of Bengaluru City 20
Annexure 1.2: Study Area. 21
Annexure 1.3: Socio-Economic Status 24
Annexure 1.4: Population and Households. 24
Annexure 1.5: Literacy Rate. 25
Annexure 1.6: Gender Status 25
Annexure 1.7: Domestic and Drinking Water Sources. 26
References. 26

2 Land Use Changes and Groundwater Overuse 29
2.1 Background. 29
2.2 Methodology. 30
2.2.1 Land Use Classes 30
2.2.2 Description of Land Use Classes 31
2.2.3 Well Inventory 32

2.3 Results 33
2.3.1 Land Use Changes 33
2.3.2 Groundwater Ovcr Exploitation 43
2.3.3 Distribution of Bore Wells as Against Land Use 52
2.3.4 Farmers' Perceptions and Interventions 57
2.3.5 Farmers' Coping Mechanisms 65
2.3.6 Adaptations to Water Scarcity: Technological Adaptation 66
2.3.7 Non-Technological Adaptation: Changing the Cropping Pattern 68
2.4 Changing Ecosystem of Jadigenahalli – Push and Pull Factors 68
2.4.1 Agroforestry 70
2.4.2 Status of Environment in Jadigenahalli 73
2.4.3 Natural Resources Stock – Land Use and Groundwater 73
2.5 Societal Responses 74
2.6 Summary 76
Annexures 77
References 91

3 Holistic Health for Well-Being 93
3.1 Background 93
3.2 Dr. Mathai's Rural Health Centre (DMRC) Soukya Foundation 98
3.3 Key Findings 99
3.3.1 Awareness Creation 99
3.3.2 Experiences and Perceptions 106
3.3.3 Challenges 110
3.4 Towards a Sustainable Approach 113
3.4.1 Youth Organisations 114
3.4.2 Improving Communication Strategies – Home Remedies in Printed Form 116
3.4.3 Influencing People on Life Style Changes and Holistic Treatment 116
3.4.4 Need to Promote Organic Farming 117
3.4.5 Need for Improved Sanitation 117
3.5 Summary 119
Annexures 120
References 120

4 Cultural Influences on Health, Traditions and Ecology 123
4.1 Background 123
4.2 Local Ecology and People's Health 128
4.2.1 Common Ailments 129
4.3 Socio-Cultural Practices and Ecological Significance 138

4.4 Sacred Trees, Temples and Ponds. 140
4.4.1 Ashwath Katte (Sacred Trees) 141
4.4.2 Ownership & Management of Sacred Trees 142
4.4.3 People Worshipping Trees 143
4.5 Temples and Natural Resources Conservation 144
4.5.1 Village Temples 144
4.5.2 Sacred Ponds 147
4.6 Summary. 150
References 151

5 Summary 155
References 160

Index 161

List of Figures

Fig. 1.1 Bangalore metropolitan region with LPAs 16
Fig. 1.2 Study area and location of villages 20
Fig. 1.3 Location of Jadigenahalli gram panchayat 22
Fig. 1.4 Villages Surveyed under Jadigenahalli gram panchayat. (Source: Google earth map, Scale 1:2 km) 23
Fig. 1.5 Annual rainfall 23
Fig. 1.6 Growth rates of Jadigenahalli in terms of HHs and population at two different points of time (%) 24
Fig. 1.7 Literacy rate 25

Fig. 2.1 Land use in Jadigenahalli during 1973 36
Fig. 2.2 Land use pattern under Jadigenahalli – 2003 36
Fig. 2.3 Land use pattern under Jadigenahalli – 2009 37
Fig. 2.4 Land use pattern under Jadigenahalli – 2013 37
Fig. 2.5 Land use pattern under Jadigenahalli – 2017 38

Plate 2.1 Brick kiln near Govindapura 38
Plate 2.2 Unoccupied residential layout near Kolathur 39
Plate 2.3 Eucalyptus plantation near Jadigenahalli (Forest Plantation) 39
Plate 2.4 Kharif crop (Ragi) near Vadigehalli 40
Plate 2.5 Grape garden near Jadigenahalli 40
Plate 2.6 Rocky area near Haralur 41
Plate 2.7 Eucalyptus plantation in agriculture land near Govindapura 41
Plate 2.8 Scrubland near Jadigenahalli 42

Fig. 2.6 Land use under Jadigenahalli GP from 1973 to 2017 48
Fig. 2.7 Variations in the extent of built-up land under Jadigenahalli GP . . 48
Fig. 2.8 Increasing trend of layouts under Jadigenahalli GP 48
Fig. 2.9 Decreasing trend of agricultural lands under Jadigenahalli GP . . . 49
Fig. 2.10 Varying pattern of agricultural gardens under Jadigenahalli GP . . 49
Fig. 2.11 Extent of eucalyptus (Commercial) plantations under Jadigenahalli GP 50

- Fig. 2.12 Distribution of wells under Jadigenahalli gram panchayat 52
- Plate 2.9 Old open well (currently not in use) near Kolathur 53
- Plate 2.10 One of the open wells having good storage near Vadigehalli..... 53
- Plate 2.11 Old water storage structure (Kalyani) near Jadigenahalli 54
- Plate 2.12 Functional bore well providing irrigation to vegetables near Jadigenahalli.. 54
- Plate 2.13 Functional bore well used for irrigation of maize near Kolathur.. 55
- Plate 2.14 Unused bore well located in Scrubland near Haralur........... 55
- Plate 2.15 Defunct bore well Located in eucalyptus plantation (earlier Crop Land) near Haralur 56
- Plate 2.16 Bore well drilled in tank bed near Kolathur.................. 56
- Fig. 2.13 Overall status of groundwater withdrawal structures under Jadigenahalli GP.................................. 57
- Fig. 2.14 Variations in the depth of bore wells under Jadigenahalli GP 57
- Fig. 2.15 Depths of bore wells in different villages under Jadigenahalli GP.. 58
- Fig. 2.16 Distribution of bore wells under Jadigenahalli gram panchayat .. 58
- Fig. 2.17 Status of bore wells under Jadigenahalli GP 59
- Fig. 2.18 Bore well depths indicated by farmers...................... 59
- Fig. 2.19 Status of bore wells across different land use classes 61
- Fig. 2.20 Land use-wise distribution of open wells.................... 61
- Fig. 2.21 Defunct open wells under Jadigenahalli GP................. 62
- Fig. 2.22 Ground water depletion: scanty rain fall 62
- Fig. 2.23 Increased number of borewells: over extraction of ground water 63
- Fig. 2.24 Respondents' views on tank degradation across the study villages 64
- Fig. 2.25 Respondents' views on deforestation across the study villages... 64
- Fig. 2.26 Livelihood impacts of groundwater depletion 65
- Fig. 2.27 Coping mechanisms and adaptations 66
- Plate 2.17 Dried and abandoned open well and bore well 67
- Plate 2.18 Cultivation of coriander (left), carrot (right) using drip irrigation 67
- Plate 2.19 Mulching done for growing rosemary, Jadigenhalli Village 68
- Plate 2.20 Transporting of eucalyptus leafs for brick industrial units....... 69
- Fig. 3.1 Awareness creation and popularity (%) 101
- Plate 3.1 Children practicing yoga in schools, Jadigenahalli village 104
- Plate 3.2 Eating healthy – promoting nutritious food habits among children by Ayush Grama Soukya Foundation.......... 105
- Fig. 3.2 New patients visiting DMRC, SOUKYA Foundation (May-Nov 2013)...................................... 106
- Fig. 3.3 Patients with different kinds of aliments (% to total)........... 107

Fig. 3.4 Perceptions about ayurveda and homeopathy 109
Fig. 3.5 Rating of ayurveda and homeopathy 109
Fig. 3.6 Number of patients visiting AYUSH Grama DMRC, Soukya Foundation (2013–14) 111
Fig. 3.7 Progress in toilet construction 118

Fig. 4.1 Common health problems 129
Fig. 4.2 People growing medicinal plants 130
Fig. 4.3 Medicinal plant usage (%) 136
Fig. 4.4 Usage of medicinal plants for different ailments (% share) 137
Fig. 4.5 Parts of medicinal plants used (%) 137
Fig. 4.6 Medicinal plant usage methods 138

Plate 4.1 Worshipping Tulsi plant 140
Plate 4.2 Worshipping sacred trees 141

Fig. 4.7 Sacred trees worship 142
Fig. 4.8 Caste groups worshipping sacred trees 143
Fig. 4.9 Reasons underlying worshiping of sacred trees 143
Fig. 4.10 People worshipping trees 144

Plate 4.3 Lamps made of Lemon (left), Lamps made of rice flour- jaggery – Ugadi festival (right) 146
Plate 4.4 Kala Byraveshwara temple in Vadgehalli 147
Plate 4.5 Sacred pond as part of the cultural heritage 148
Plate 4.6 Mango leaves tied to the entrance of house(right) and the temple entrance(left side) 149

List of Tables

Table 1.1 Proportion of members who reported to have received any training to carry out their principal activity by type of activity (in percentage) 10
Table 1.2 Composition of Jadigenahalli gram panchayat 21
Table 1.3 Types of housing under Jadigenahalli gram panchayat (%) 22

Table 2.1 Land use/land cover schema adopted for Jadigenahalli 31
Table 2.2 Land use/land cover over the years in Jadigenahalli area in acres 33
Table 2.3 Land use pattern under Jadigenahalli gram panchayat: 1973–2017. 34
Table 2.4 Current land use pattern (2017) under Jadigenahalli GP – % distribution of classes across villages 35
Table 2.5 Groundwater withdrawal structures under Jadigenahalli gram panchayat 51
Table 2.6 Location of wells in relation to land use 60
Table 2.7 Distribution of defunct open wells under Jadigenahalli GP 61
Table 2.8 Status of bore wells as of 2014 62
Table 2.9 Investment on functional bore-wells (Rs. Lakh) 63
Table 2.10 Distribution of eucalyptus plantations under Jadigenahalli GP 71
Table 2.11 Village-wise and land use-wise distribution of wells under Jadigenahalli GP 72

Annex 2.1 Current land use pattern (2017) under Jadigenahalli GP 78
Annex 2.2 Land use pattern (2013) under Jadigenahalli GP 79
Annex 2.3 Land use pattern (2009) under Jadigenahalli GP 80
Annex 2.4 Land use pattern (2003) under Jadigenahalli GP 81
Annex 2.5 Land use pattern (1973) under Jadigenahalli GP 82
Annex 2.6 Depth (variation) of bore wells under Jadigenahalli GP. 83
Annex 2.7 Village-wise and land use-GP wise distribution of wells under Jadigenahalli GP 84

Annex 2.8 Bore well depth and land use relationship under Jadigenahalli gram panchayat 86
Annex 2.9 Distribution and status of borewells under Jadigenahalli Gram Panchayat 88
Annex 2.10 Bore well depth and land use relationship under Jadigenahalli gram panchayat 90

Table 3.1 Activities and status. (Source: Soukya Foundation Documents) 100
Table 3.2 People's opinion about DMRC, Soukya Foundation according to ASHA worker 111
Annex 3.1 NGOs involved in Ayush Grama 120

Table 4.1 Medicinal plant usage and practices 128
Table 4.2 Medicinal plants grown in Jadigenhalli gram panchayat 131
Table 4.3 Medicinal plants grown in percentage 135
Table 4.4 Religious significance and belief systems attached to plants/trees 139
Table 4.5 Temples and belief systems 145

Chapter 1
Introduction

Abstract This chapter introduces the challenges of peri-urbanization and its implications with references to climate change. It also introduces the key questions that will be answered in the consecutive chapters on land use change and implications, public-private partnership governance model of health care under the AYUSH program and role of culture in protecting environment and health traditions combating climate change. An introduction to the study area, the semi-arid region of Bengaluru, Jadigenahalli, and methodology is also discussed in-depth in this chapter.

Keywords Peri-urban · Climate change · Jadigenahalli gram panchayat

1.1 Why This Book?

The rural urban linkages, their interdependence and interrelations have always been prevailing with far-reaching impacts on ecosystems and development at large. However, in the recent times, newer forms of these relations have been observed, as to state 'periphery areas' are hazy to distinguish as rural or urban; more so in the present context of a rampant urban expansion. Urbanization has been significantly high, especially since the Industrial Revolution and given the projections, the proportion of people living in cities and peri-urban areas will reach to 66% by 2050, with a majority of the million-plus cities and megacities concentrated in India and China (UN 2018). In the Indian context, 2011 census indicated, for the first time in seven decades, the urban population recording a higher growth than the rural population. Rapid urbanization and related changes in land use patterns have led to transformations that are affecting the ecosystems and urban environs at large. As a resultant effect, predominantly, we observe prominent land use changes and associated economic activities in the vicinity of rural regions.

In peri-urban contexts, rural and urban areas are juxtaposed with increased landscape changes being witnessed, caused mainly by anthropogenic activities. These areas are being termed as 'peri-urban' areas by experts (Adell 1999; Alen 2003; Davis 2004). However, debates over conceptualizing the peri-urban' have also been diverse (Adell 1999). This may be attributed to the 'Changes' observed, while

S. Manasi, K. V. Raju, *Coping Mechanisms for Climate Change in Peri-Urban Areas*, https://doi.org/10.1007/978-3-030-18517-6_1

evidences have shown that it has been diverse across peripheries (Adell 1999). Further, it is also important to distinguish that the changes are not just restricted to cities; rather, it is different across the boundaries/peripheries of the cities as well, making it more complex and difficult to define the nature of these peri-urban areas (Adell 1999). We discuss the conceptualization of 'peri-urban' by various experts at a later stage. However, we would like to indicate that several 'Peri-Urban' linked deliberations have brought to the fore some of the challenges and concerns facing in peri-urban areas, their severity and extent, that need to be addressed. These discussions have also necessitated a nuanced understanding of and research in 'peri-urban' contexts.

In South Asia, peri-urban areas are expanding at a much faster pace than many core cities and towns. In the Indian context, assessments studies of the dynamics driving peri-urban changes in the current scenario appear inadequate. A rapid expansion of city limits can be seen as the new urban reality of India – a quintessential feature of contemporary urbanization. This expansion of peri-urban areas in India is exerting increased pressures on infrastructure provision (Singh 2018); besides leading to an imbalanced development and the associate challenges- water, sanitation, waste disposal, drainage, roads, housing, parks, schools, hospitals, public transport and others.

Globally, climate change linked concerns are the most focused and discussed issue. 'Peri urban areas' are vulnerable to 'Climate Change' risks given their positioning, since they belong neither to the rural nor urban arena. Evidences point to an extensive damage to the provision of ecosystem services in peri urban environments. Hence, peri urban areas are prone to environmental impacts due to climate change risks. Further, urban and peri-urban environments are characterized by an extensive land use and ecological diversity (Douglas 2008). Besides, 'Climate change' is also a matter for concern and a challenge to regional and urban planners, as traditional challenges, viz., governance, funding, rapid growth of geographical areas and population, increasing demand for infrastructure (transportation, water, sanitation, energy), expanding social services (education), implications like pollution types and extent, slums and associated aspects, are exacerbated by the need to more fully accommodate the direct and indirect impacts of climate change and ecosystem loss in the planning process (Clark 2009). Such unplanned growth, coupled with environmental changes, can certainly pose a severe challenge to sustainable development. At large, there are several reasons for not integrating fringe areas, specifically urban ecosystems and climate change, into the framework of traditional urban planning. Major reasons are attributed to the urbanization process, environment and ecology of the fringe areas, lack of adequate information on the 'value' and 'place' of ecosystems in urban planning and the interaction of climate change and cities (Clark 2009).

Given this, backdrop, a global city like Bengaluru has been witnessing a wide range of changes. The local residents (including the authors) have experienced the city expansion and its magnitude and have for sure felt not only overwhelmed, but also petrified of several challenges, both for present and future. Our interest in the present issue increased with several readings (T.V. Ramachandra et.al 2012; Narain et.al 2013) including research studies that we closely involved with over the years (Raju et al. 2003; Manasi et al. 2006, Manasi et al. 2014; Sridhar et al. 2016). Like other peri-urban regions, Bengaluru city also has undergone changes that may be similar to other urban areas. However, there are some unique features specific only to Bengaluru with varied implications. Climate change dimension of peri urban areas is also linked to ecosystems of these regions. Peri-urban areas encompass a combination of diverse land-uses, consisting of paving surfaces, soils and vegetation structured in complex formations, with forests and agricultural fields forming a vital part. Patterns of urban expansion and growth differ extensively by region, with diverse implications for sustainability. Given the increasing concentration of people and the extent of these areas, together with climate change projections, urban and peri-urban sustainability and safety are a growing concern. Keeping those obvious features in perspective, we wanted to understand at the micro level the dynamics, vulnerabilities and coping mechanisms of peri-urban regions. Since the peri-urban regions of Bengaluru fall under the semi-arid zone, we chose Jadigenahalli Village Council (*Grama Panchayat),* for an indepth exploration as part of understanding and analyzing some of the pertinent questions related to this peri-urban region, specifically coping mechanisms in the context of climate change. The present study is intended to help understand climate change impacts, vulnerabilities, and coping mechanisms of semi-arid regions. The questions include -

1. To what extent has the growth in Bengaluru impacted land use change and its implications for agriculture and associated livelihoods in the peri urban regions? What are the causes, implications, its extent and coping mechanisms?
2. How has the governance structures fared in addressing the challenges of health and sanitation concerns in peri-urban regions?
3. What is the status of and issues related to 'water' in specific? What has been the role of culture-led informal institutions in protecting the environment? (given the fact that Indian villages are an embodiment of nature and culture based society). To what extent have they acted as coping climate change implications?

With this contextual framework, our book brings out the dynamics of underlying resource management in the peri-urban interface involving climate change consequences. Besides capturing the situation, by understanding the processes and functions of these regions, we identify sustainable and feasible adaptation strategies, address ways in which coping mechanisms are adapted and ways in which, capacities could be built into the system to further strengthen the resilience of mechanisms and sustainable development in semi arid regions.

We have included understanding the ways and assessment of how communities, government and other service providers are responding to challenges and opportunities in these rapidly changing landscapes, given the fact that, without an appropriate planning, urbanization can cause a considerable damage to the natural environment. Since peri-urban areas are geographical environments near a city or on the outskirts of a city and are essentially transitional areas changing from rural to peri-urban before possibly becoming urban. Over time, the characteristics of these areas keep changing. Further, considering that these areas expand to encompassing villages, towns and urban areas originally outside of their boundaries, these peri-urban areas tend to urbanize rapidly. Also, the growth of these areas is often unplanned, as many of them do not fall under a given city's official jurisdiction and hence, lack critical infrastructure facilities and governance systems.

The resultant conditions affect the traditional livelihoods, victimising the residents of peri-urban areas and also make them vulnerable to the adverse impacts of climate change. These crucial areas with land cover changes result in the transformation of hydrological, ecological, and geomorphologic as socio-economic systems are time and again neglected by both rural and urban administrators (Narain 2013). We will capture the issues keeping all these changes in perspective. In the light of these characteristics, the book documents at length, the challenges encountered resulting from impacts on the local and regional water resources and their implications. Besides, the book also brings out the experiences related to environmental management by local institutions, government interventions that act as catalysts in promoting community based autonomous adaptation strategies besides edifying the neglected part of urban development processes and the physical, social and economic rural-urban dynamics.

The analysis enables a better understanding of how improved adaptation mechanisms in peri-urban areas are important and the ways in which anthropogenic pressures play a major role in impacting ecology, role and interventions made by local environmental and urban policies. The narrative also indicates how varied strategies are adopted for coping and their contribution to the improvement of local strategies. Further, the book identifies the need for action plans in urban areas, which may bring about positive environmental changes. It is also that these peri-urban areas have huge prospects, as they will translate to urban areas. There is a need to identify innovative approaches and initiatives that are essential to transform these urban fringes and to help design strategies for sustainable development of peri-urban areas.

1.2 Peri-Urban Areas

The word 'peri-urban' is indeed a confusing term with many different conceptual connotations. Nevertheless, the word 'peri-urban' is mainly used in three different ways- namely, as a place, a process or a concept (Narain and Nischal 2007). Peri-urban areas refer to transition or interaction zones, where there is a juxtaposition of urban and rural activities and landscape features are subject to rapid modifications due to human activities (Douglas 2008; Dutta 2012). Peri-urban areas are transitional zones with a significant governance ambiguity. These areas lie beyond municipal jurisdiction and public services viz. water, sanitation and waste disposal are not provided or at times provided in a limited manner by rural or urban authorities. Although these peri-urban fringes are not overlooked in the national, regional and urban planning activities, certain issues have not got the attention required, such as the rapid expansion of urban centers into their fringe zones viz. peri-urban, peri-agricultural, agricultural and undeveloped land and the impact of urbanization on ecosystem sustainability. Both of these aspects are crucial to sustainable urban development, but, both of these are not understood well, more so with respect to urban planning and development (Raghunath 2016; Clark 2009).

Mycoo (2006) characterizes the peri-urban interface as a place that involves many processes that impact people; it is also a zone in which the processes of social exclusion is experienced by many inhabitants of informal settlements who are poorly served by infrastructure and services. There is a growing recognition that sustainability of both cities and rural areas is affected by dynamic and changing flows of commodities, capital, natural resources, people and pollution within the peri-urban interface. This zone is a place of competing interests, which lacks approaches that strike a balance required to ameliorate poverty, protect the environment, and maximize the productivity of human and natural resources, to draw synergy from the urban and rural relationships. Peri-urban settlements reflect persistent neglect, conflict and competition.

Narain (2007) reiterates that 'peri-urban' should not be used in its narrow, geographical sense rather as an analytic construct to study the relationships between urban and rural activities, processes and institutions. This is based on his argument that the notion of peri-urban is most widely understood conceptualization and he identifies two main approaches to defining the peri-urban interface – (i) spatially, as a transitional zone around a city;and (ii) as a zone of intense interactions, flows and linkages between urban and rural areas. When used in this sense, 'peri-urban' refers to rural fringe areas surrounding cities that bear the 'spillover' effect of urban expansion. These areas provide the much needed land and water resources for urban expansion besides serving as containers of urban wastes. Besides this, peri-urban areas are perceived differently by policy makers, academics, the powerful and the powerless. There is a poor understanding of the peri-urban environment, the diverse livelihoods of people experiencing different degrees of urbanization, and the perceptions and priorities of peri-urban villagers. Hence, the planning decisions are inevitably flawed. Since there is no clear jurisdiction, 'organized irresponsibility' has emerged allowing powerful actors to benefit from the lack of regulation and constant impermanence. Meanwhile, the marginalised and disenfranchised groups lack access to basic health, water and sanitation services. Often, their rights to land, health and water are taken away. Besides, their livelihoods are frequently declared 'illegal' by the state and the judiciary.

Peri-urban areas are characterized by great heterogeneity and rapid changes in land use. Furthermore, population composition changes as peri-urban areas offer attractive residential alternatives to city centres or more remote locations. The dynamic processes leave peri-urban areas in an in-between situation, neither city nor countryside and home to a range of functions, spanning from agricultural production to residential and recreational areas (Busck et al. 2006). Peri-urban areas do not fall under the purview of a city's administrative unit but, adjacent to it and are transition zones between urbanized receptacles and rural agricultural vicinities. Unfortunately, the loss of agricultural and undeveloped lands, unauthorized urban development and industrial operations, environmental degradation and overall a significant alteration of critical ecosystems are serious and ever increasing problems faced by fringe zones (Clark 2009). The peri-urban regions are also susceptible to resource constraints, access to services like water, energy, transport and housing get increasingly distorted with changes in land use. Besides, unplanned growth impacts environmental changes and poses threats and challenges to sustainable development. Ecologically, the transformation of urban fringe zones into alternative uses is potentially the most severe facet of the rural-urban transformation, the implications of which are in many situations irremediable, have broad complex linkages throughout the urban environment and are presently largely unrecognized (Clark 2009).

Due to a rapid urban growth, city periphery is primarily subject to multiple transformations: physical, morphological, socio-demographic, cultural, economic and functional. These transformations make this area experience a high spatial uncertainty resulting in undesirable, complicated land use land cover (LULC) patterns. As the peri-urban areas emerge at a faster pace, there is a need for protecting common property resources that are otherwise diverted to other activities and purposes (Narain 2009). The present urbanization process is taking place at the expense of peri-urban ecosystems in that it draws heavily on the former for the creation of differentiated urban structures, which have, in the process, led to serious problems like depleting ground water resources, changing cropping patterns, emergence of factories, impacting the livelihood of rural people in varied forms.

1.3 Evolution of Peri-Urban Areas Around Indian Cities

Over the last century, India's landscape has urbanized rapidly. The urban population has grown 14 times since 1901 and is expected that 54% of India's population will be urban by 2050 (ACCCRN). Most of the growth is observed in large villages or small to medium sized cities (IIHS 2011) resulting in vast 'peri-urban landscapes. By 2050, three quarters of the world's population will live in cities and in Indian situation. With Prime Minister Modi's Smart City Mission, 20 cities of the 100 planned are identified; hence, a new development process of these secondary cities will likely act as an additional driver of land use change between satellite towns and the close primary cities even as areas on the peripheries of cities are continuously changing and expanding. Besides spatial terms, the peri-urban interface is increasingly recognized in terms of dynamic flows of commodities, natural resources, people and pollution. Although often seen as a transition zone, peri-urban areas are expanding prominently.

Currently, there are no appraisals regarding the dynamics driving peri-urban change in India. There is significant gap in the literature with respect to the conceptualization of peri -urban areas, defined and conceptualized in different ways. 'Peri-urban areas are inherently dynamic spaces experiencing significant changes to landscapes and livelihoods (Bahadur 2016). Bahadur (2016), notes that 70% of the urban infrastructure required by 2030 is yet to be built and more likely to be built in peri-urban areas. This also indicates that there will be a further transformation of the rural hinterlands across the country, with differential patterns of peri-urbanization. Murali from IDRC (2014), views peri-urban areas as the interface between town and country with urban and rural uses mixing and sometimes clashing. Dr. Sumetee Pahwa Gajjar (2014) views peri-urban in the context of vulnerability and social differentiation perspective and while stressing the need for identifying peri-urban areas as zones of contestation that display a dense overlap of geographical, socio-economic, political and environmental dimensions of marginality. Amit Mitra (2015), an independent researcher viewed peri-urban areas as the unwanted child of rural and urban.

Datta (2014), proclaims that, peri-urban areas across the country may not be the same across varied dimensions of urbanity. In the Indian context, on the administrative front, there is a lack of clarity about the notion of 'peri-urban', but its importance has been recognized in terms of resilience building of urban centres and survival of the poor. With the population increasing, in peri-urban areas of cities, the stress of managing and financing infrastructure services in peri-urban areas is also getting accentuated. Increased inequalities in urban areas have created two types of urban structures within Indian urban areas. (i) Supporting a lifestyle equivalent to that of any developed country; (ii) Poor quality of life strongly exhibited in the unhygienic and dilapidated conditions. Simultaneously, these two parallel situations prevail. In peri-urban areas, there is a growing demand for increased quality of infrastructure by the relatively higher- income urbanites, while there is a lack of even basic services for the poor. It is evident from most of the cases that a substantial proportion of unskilled labour is losing hold on their land because of changing land uses in peri-urban areas. They are mainly drawn into the unorganized sector primarily in low -productivity self-employment.

Projects like JNNURM, aimed at improving access to basic services and support an enabling environment for economic growth have been implemented within urban areas by urban local bodies and not in peri-urban areas, as they in general are under Panchayats whose major function is to plan for economic development and ensure social justice and to implement these plans, as per the 73rd amendment of the Constitution. However, the Panchayats are unable to meet the overpowering demand for services in these areas, due to limitations of finance and lack of capacity for planning and managing the required infrastructure. For example, most Panchayats lack capacity to deal with solid waste management, the most pressing issue in peri-urban areas. These areas receive not only solid waste from core cities but also their own generated waste to be managed. Usha (2016) describes peri-urban as transitional zones characterized by a significant governance ambiguity. She reiterates the concern that these areas lie beyond municipal jurisdiction as a result of which public services such as water, sanitation and waste disposal are not provided by rural or urban authorities.

It is also important to note that rapid urbanization has increased the wealth and purchasing power of India's urban population. Besides, urban development has also brought about changes in the land use affecting traditional industries, agriculture, fishing and forestry, which are mostly common in peri-urban regions thereby leading to newer challenges. Furthermore, climate change and frequent extreme weather related events have been observed posing further problems to both urbanizing communities and climate -dependent livelihoods like in fringe areas depleting water levels affecting agriculture and with changing sea levels and temperatures. Two types of transformations happen in peri -urban areas i.e., 'built environment' and 'lived environment'. It has been highlighted that changes happen in the occupation and social structures of residents due land use transformation with the dwellers taking up informal occupations, due to lack of skills and loss of their land to developers in the upcoming peri-urban areas. There is a change observed in the consumption of goods and a contest for natural resources due to sudden changes in the ownership and a sense of belonging to their lands within a resultant transformation of the social fabric of the area. Women are seen seeking new unskilled livelihood opportunities in the organized sectors like IT industries. Besides, a rise in crime and corruption in peri-urban areas can also be witnessed. According to Land acquisition Act, the acquired land cannot be sold as freehold, but leased and this results in illegal ways of support sale of land. Building permissions are currently under village councils *(Panchayat)*, therefore, there is an obvious resistance to changing these areas into 'urban'. Hence, the role of urban planners in viewing peri urban areas as beyond geographic units and should be looked as a process. A different analytical construct and innovative planning are required to look into urban-rural issues (Ratoola Kundu 2014).

1.3.1 Peri-Urban Ecosystems

'Peri-urban' in the ecosystem context need to be seen as a heterogeneous mosaic of natural ecosystems, productive or agro -ecosystems and urban ecosystems affected by the material and energy flows demanded by both urban and rural systems. Since there exist lacunae in clear terms of cut policies and plans, peri-urban areas face uncertain land tenures accompanied by inferior infrastructure (Prakash 2012). Hence, this results in environmental and natural resource problems beyond the scope of urban or rural governments. It is also foreseen that if the same trend continues, the situation across peri-urban regions will possibly worsen with impacts of climate change. Hence, the need to understand the physical, social, economic and institutional aspects of peri-urban areas based on innovative approaches is important besides evolving protective measures to maintain their provisioning, regulatory, supporting and cultural services (Prakash 2012; Narain 2010).

The peri-urban ecosystems, upon which a large proportion of the population depends for environmental services, are getting increasingly affected by a rapid urbanization process, with uncertainties and newer challenges looming over the populace. More importantly, it is the peri-urban interface around larger cities or urban centres which has been driving the processes of urbanization with significant environmental impacts that can be observed across various arenas and levels, rendering the dynamics underlying the rural-urban ecosystems more complex and complicated. Peri-urban areas are observed as inherently dynamic spaces experiencing significant changes in landscapes and livelihoods (Bahadur 2016). Besides providing ecosystem services to the cities, peri-urban areas render carbon sequestration and climate regulation, water and air purifications, pest and disease control services. In the context of climate change, the resilience of peri-urban areas to local change and environmental degradation needs to be analyzed to understand the challenges encountered and their implications.

With the increasing demands for infrastructure and pressures on natural resource systems, many peri-urban dwellers, particularly in the context of urban demand for natural resources for non-agricultural uses, are affected by urban-generated wastes. Cities also exert a heavy pressure on their surrounding regions by depleting fresh water resources. Most urban wastes end up on the outskirts of city with solid wastes being disposed off on peri-urban land sites, while liquid wastes are let off into rivers and water bodies located close by. Besides, peri-urban areas are also getting increasingly affected by urban air pollution. Land markets are another important component subject to competitive pressures, as urban centres expand with speculation, increasing land prices. Under such conditions, low-income groups like small and marginal farmers or residents of informal settlements either benefit from these changes or end up losing access to land, depending on the land rights system prevailing in a given region. Thus, the traditional relationships between the village and the city, the farm and the factory have undergone significant changes besides becoming increasingly complex in the process of globalization. A glimpse into these areas provides an array of pictures of the rural-urban landscape with several issues characterising the continuing transformation. Further, there are various dimensions of development, many of which may not have been documented in detail, but are taken note of across various discussions.

Climate change directly affects food and nutrition security, undermining the current efforts towards addressing under nutrition, one of the world's most serious but least addressed socio-economic and health problems. A combination of nutrition-sensitive adaptation and mitigation measures and nutrition-smart investments increase policy coherence, while institutional and cross sectoral collaborations can address threats to food and nutrition security from climate change. A rights based approach engages the peri-urban stakeholders most vulnerable and affected by climate impacts as active participants in this process (Shaikh and Jeevanjyoti 2013).

Another important aspect relates to the changing farming systems and the level of labour participation. Although peri-urban agriculture can be profitable, the small and marginal farmers may get eliminated in the process with large farmers investing in agricultural intensification. Equally important is the revival of peri-urban agriculture for nutritional and food security of urban residents; the socio-economic marginalized populace living in the peri-urban areas also face nutritional security (FAO 2011). In such a scenario, agricultural wage labour becomes more significant than small-scale farming and also for attracting migrant workers. A critical dimension is residents of peri-urban areas benefit from job opportunities in the urban areas and emerges as entrepreneurs and invests in various business options, including real estate, taxi, vehicle rentals, small shops, food grains and vegetable trade, goods transport and construction. Various migration forms have also influenced lifestyle changes among the rural populace. A large number of migrants find job opportunities in urban areas as unskilled labourers (Srivastava 2003). Newer forms of informal employment in shopping malls and parking sections as security guards, elevator operators, and helpers in restaurants and others, both skilled and unskilled workers can also be observed.

Some of these influences have resulted in both positive and negative implications. For instance, children opting for private schools, which was 2% in 1980 increased to 21% by 2014, indicating that the rural poor incur a huge expenditure on children's education (Gupta 2014). On the other hand, negligence of traditional management practices, informal institutions, changing cropping patterns due to depleted groundwater levels have affected livelihoods, increased consumerism and associated changes. At the village level, there has been a shift towards non-farm employment all over the country. The National Sample Survey (2009–10) shows that the percentage of non-agricultural households has increased from a pre-existing high of 31.9% in 1993–94 to 42.5% in 2009–10 (National Sample Survey). The recent survey by the National Bank for Agriculture and Rural Development (NABARD, 2018) on financial inclusion in rural areas, reveals startling numbers about reduced dependency on crop-based income; across all households in rural areas, wage labour contributes 43% of the total income, 24% service from (government or private), cultivation 19%, and livestock sources 4% (See Table 1.1).

The Census also indicates that factories and workshops have a strong presence in rural areas. The Micro Small and Medium Enterprises (MSME) industries Census indicates that rural India accounts for 45.2% of the registered units, while 60.2% are

Table 1.1 Proportion of members who reported to have received any training to carry out their principal activity by type of activity (in percentage)

Characteristics	Agricultural households	Non-agricultural households	All households
1	2	3	4
Overall			
	By type of usual activity		
Self employed / own account worker	5.3	13.7	7.5
Employer	18.0	18.1	18.0
Unpaid family worker	3.8	2.8	3.5
Regular salaried / wage employee	28.1	32.4	30.9
Casual wage labour in public works (other than MNREGA)	4.9	5.0	5.0
Wage labour-MNREGA	8.8	3.7	5.6
Wage labour- other works	12.7	17.0	15.6
Does not work, but seeking/ is available for work	8.3	5.0	6.5
Attends to domestic duties with sewing/weaving etc.	5.6	6.5	6.1

Base = All Household Members
Source: NABARD All India Rural Financial Inclusion Survey 2016–17

unregistered ones (Gupta 2014). As the real estate prices are low relatively, several factories are establishing operations in peri-urban areas.

Given this scenario, it is critical that we have a better understanding of the linkages between urban and peri-urban areas by addressing ecosystem services that are vital to the livelihoods of people besides their resilience to climate change shocks. To restore ecosystems, there is a need for approaches that are multi-faceted and multi-dimensional by involving stakeholders across all levels. Besides, reviving of ecosystems also should be inclusive of people's development needs, particularly health and education (Mitra Amit et al. 2015).

1.3.2 Vulnerability in Peri-Urban Settlements

Indian cities are characterized by an extensive peri-urban development in the context of urban centres expanding into rural areas in a phenomenon referred to as an urban sprawl. The resulting settlement patterns may include the following features.

Settlement Patterns

(i) Affluent residential areas – Affluent residential areas are usually that result of a gradual organic urban growth, more often on land that has a high agricultural potential or scenic value and is also the result of lifestyle choices of high- networth individuals and families. Such residential areas are often beneficial for high end property developers and there are strong market pressures to exclude low-cost housing or social housing developments on this land. These areas typically have a reliable access to basic services and high consumption patterns of water and electricity. They typically have a high adaptive capacity, but may support patterns of water and energy consumption that are not environmentally sustainable.

(ii) Gap, low-cost and social housing estates – this includes many new housing estates dating from the post 1994 era often on degraded land with a low market value, leasing of informal or semi-formal structures in backyards of low-cost or gap-housing, and these households are likely to be highly insccure in terms of tenure and other indicators of social vulnerability. Price has played a defining role in the design and construction of low cost housing and these structures are generally not 'climate proof'. They may be structurally vulnerable to extreme weather, poorly insulated against temperature extremes and prone to leaks and flooding during heavy rainfall. Equally important, in the design and implementation of low-cost or gap-housing estates an inadequate attention is often paid to accessibility of public services, community facilities, recreational spaces, urban aesthetics and social cohesion.

(iii) Informal settlements – Informal settlements are particularly vulnerable to climate change for several reasons. By nature, they are unplanned, with no extensive service or infrastructure coverage. They feature construction using materials of diverse origin and quality and not always following accepted techniques. These houses rarely comply with official safety standards and there are no controls in place (Hardoy and Pandelia 2009). The majority of vacant land occupied in this way is on the urban periphery and is often located in areas that are unsustainable for human settlement due to local topographical features such as unstable soils, wetlands and flood risks. A significant percentage of the backlog in service delivery is associated with informal settlements. Informal settlements often have high population densities with residents being highly socially vulnerable to climate change.

(iv) Mixed settlements – These combine features of urban settlements and rural settlements. They are typically not as densely settled as urban areas, but are too densely settled to support intensive commercial agriculture. They may include traditional, formal and informal housing types with access to basic services often being patchy and unreliable. This settlement type exists only on the perimeter of some urban settlements, but also is a characteristic settlement pattern in the former homelands. It is further characterized by settlement layouts that are not formally planned, which can complicate service delivery and uncertainty in relation to land tenure associated with the administration of communal lands. In general, poorly built households located in peri-urban settlements are likely to experience a heightened vulnerability to climate change not only as an intrinsic consequence of socio- economic demographics, but also due to the distances wage earners need to travel, inadequate access to basic services, insecurity of tenure, and physical vulnerabilities of informal unplanned or poorly planned housing.

1.3.3 Water Crisis and Insecurity in Peri-Urban Contexts

The United Nations Conference on Sustainable Development 2012 (Rio + 20) states that water is a fundamental resource and one of the main themes discussed in the context of Sustainable Development and Climate Change. Since water is greatly linked to other natural resources like food, energy, it is the most vital resource for survival. During the Bonn Conference (2011), it was indicated that there was a need for analyzing the interrelation between the three resources – 'water, energy and food security nexus platform.' A research study on Climate Policy, Conflicts and Cooperation in Peri-Urban South Asia has demonstrated that the processes of climate change and urbanization are creating growing problems of water security in the peri-urban areas of the rapidly expanding cities of Khulna (Bangladesh), Kathmandu (Nepal) and Hyderabad and Gurgaon (India). They have also indicated that there is a critical knowledge gap that prevails in measuring their direct and indirect effects on the lives of the poor and vulnerable, and their implications for the incidence of conflict and the emergence of new forms of cooperation that are more sensitive to issues of water security for water insecure, poor and marginalized groups.

The consortium also highlighted another critical gap as devising innovative ways of using new knowledge and insights for capacity building, to support sustainable, climate smart and pro-poor development trajectories that focus on the nexus of water security, climate change and urbanization.

In the context of India, a major proportion of India's population exist in water-stressed regions. With population growth, increasing demand and inefficient usage, overexploitation of groundwater, pollution and climate change have hastened water access issues. Case studies from peri-urban areas of Gurgaon and Hyderabad have highlighted urbanization's impact on the lives of people and water sources in the regions. They also address water conflicts in the peri-urban areas of Delhi, as part of exploring technological and environmental sustainability challenges in these regions. Hence, it is considered important to understand the specific social, economic and institutional characteristics of peri-urban areas and to document more closely the changing nature of these regions (Prakash et al. 2011).

1.3.4 Peri-Urban Changes and Concerns Over Food and Nutrition

Urban and peri-urban agriculture is recognized to contribute to urban food security, household income and environmental management (Mkwambisi 2009; Mutonodzo 2009; Satterthwaite et al 2007). Agriculture is also a key part of the socio-ecological systems of peri urban areas. Ecosystem services offered by peri-urban areas to urban systems include, among others, sustainability of rural food production – food security, 'functionality' of ecosystem services, innovation in farming practices and so on. Although there is evidence to show that urban governance institutions have recognized the role of urban and peri-urban agriculture in improving urban food security and enhancing nutritional and livelihoods security, there is still a need for recognition in policy and practice. On the one hand, it is obviously seen that there is an increasing demand for affordable fresh produce for meeting the needs of urban populations, considering that the peri-urban interface is often perceived as a temporary 'belt' on the city fringes, and the decline of agriculture is often seen as inevitable to make way for urban development (Marshall et al. 2009). In this context, the argument is to bring about a fundamental change in the mindset to prevent land use changes and unregulated construction activities in peri-urban areas. There is a need for a multidisciplinary and integrated administrative land-use planning that recognizes the agricultural ecosystems as the core component of the urban and peri-urban natural resource systems (Marshall et al. 2009; Pothukuchi and Kaufman 1999). There is also a need for the promotion of good agricultural practices, and the services rendered by these areas – provision of food, micro climate mediation, carbon sequestration, nutritional recycling and recreational opportunities from multi-functional urban green spaces as part of ensuring and sustaining the livelihoods of the poor. Besides, these good agricultural practices underscore the need identifying the potential human and environmental health risks individually on a case or city basis, and develop and implement evidence-based policies with multiple stakeholders involved as part of eliminating these risks or reducing them to acceptable levels. There is evidence to show that in different parts of the world, peri-urban agriculture has been promoted and adopted as part of adapting to changing climate (Lwasa et al. 2009; De Zeeuw et al. 2011; Roberts et al. 2011).

1.3.5 Threats to Peri-Urban Forestry

Urban and peri-urban forestry is the practice of managing forests, groups of trees and individual trees in and around the urban areas to maximize their economic, livelihood, social, cultural, environmental and biodiversity values. Urban and peri-urban forestry can serve several purposes by covering natural and planted forests and trees, forests maintained in watersheds or dry lands, forests and wooded areas, green spaces and street trees as well as trees in urban/peri urban gardens. Maintaining the land use pattern beyond the city edge helps in adapt to flooding in varied ways, productive greening strategies (fruit trees, herbal shrubs, high value vegetables) by increasing water infiltration and a reduction in the time lag for flood occurrence on hill slopes and in valleys. Besides, increased tree plantation in peri-urban areas also has adaptation potential to heat waves by moderating microclimates. During summer, temperatures tend to be high, with productive greening around the peripheries of cites, temperatures get reduced and the effects of urban heat island.

1.4 Urbanization in Peri-Urban Areas

1.4.1 Urbanisation – Influence of Development Process

As discussed extensively earlier, as an area gets urbanized, the process of urbanization affects the livelihoods of people in the peri-urban areas. The process benefits a few and renders others vulnerable. Economic development policies focused on growth tend to subsidize lands in the peri-urban areas for the establishment of industrial parks or zones, leading to a potential loss of prime agricultural land or an inefficient land use and the leaving of land unused with future expectations, in the process.

Bangalore city had shown the signs of a faster growth during the 1980s. With the purpose of planning, coordinating and supervising proper and orderly development of areas around Bangalore, Government of Karnataka had constituted an autonomous body called Bangalore Metropolitan Region Development Authority (BMRDA) in 1985 under the BMRDA Act 1985. BMRDA has under its jurisdiction Bangalore Urban, Bangalore Rural and Ramanagar districts with an area of 8005 sq.kms. In accordance with BMRDA Act 1985, A Structure Plan was prepared for

Bangalore. The structure plan accounting the resources in the region, described in outline a mixed development strategy for land designation, delineation, reserving and acquisition. It highlighted the policy imperatives to encourage private sector participation in the provision of infrastructure, adoption of a better levy approach to land conversion and designation of nature preserves. While the regulatory and planning functions of BMRDA brought orderliness to the urbanization process, it had sent vibes across the region of 8005 Sq.kms about the scope for land conversion and scattered housing developments and industrial establishments.

BMRDA has further categorized the Bangalore Metropolitan Region (BMR) into Area Planning Zones (APZ's), which are the areas where urban development is permitted subject to certain regulations. The Interstitial Zones are the areas lying between APZs where urban activities are restricted with more emphasis given to environmental – issues like conservation of forest area, agriculture etc. For planned urban growth Local planning areas under Karnataka Town & Country Planning Act-1961, it had declared 8 Planning Authorities (Fig. 1.1) and one of such relevant to Jadiganahalli is Hosakote Planning Authority (HPA).

The expansion of Bengaluru city with an addition of 8 Urban Local Bodies (ULBS) and 110 villages around the city to form Bruhat Bengaluru Mahanagara Palike (BBMP) illustrates the urbanization trend and its spread into peri-urban areas of the city. The two ULBs developed to be satellite towns (Yelahanka CMC and Kengeri TMC) have now been merged with BBMP. Apart from urbanization, the investments leading to creation of jobs and market also influence the land use changes in the peri-urban area.

Hosakote was earlier characterized by large land parcels stretching out on either side of Old Madras Road with the presence of some industrial units and in the recent years, it has picked up pace in residential as well as mixed use developments. Proximity to Whitefield as well as Krishna Raj Puram (KR Puram) and the improvements to the national highway (four lane) have led to an increased real estate activity in this belt. With low-land availability and soaring prices within the core city (Bangalore) limits, Hoskote has seen increased interest from developers in the recent past. Hoskote has witnessed a major residential activity against the backdrop of relatively lower land values, good connectivity to the new airport and proximity to established Information Technology hubs of Whitefield, Outer Ring Road and Old Madras Road. Although Hosakote is located around 33.2 km off from the central part of Bangalore, its location on the NH-4 as well as its closeness to the major commercial hub of Whitefield is one of the major pull factors of development. It is well connected to the city through a good road (bus) network and also has two railway stations close by at K R Puram and Whitefield. Of late, some areas in this region, located closer towards the city, have been witnessing a number of high-rise residential buildings from leading builders.

The industrial areas around Hosakote are also creating a demand for residential options. The industrial hub with around 200 industrial units, housing auto spare parts, chassis works as well as warehousing and logistics units and the facilities of major auto manufacturers are major job options. Hosakote being well-connected to

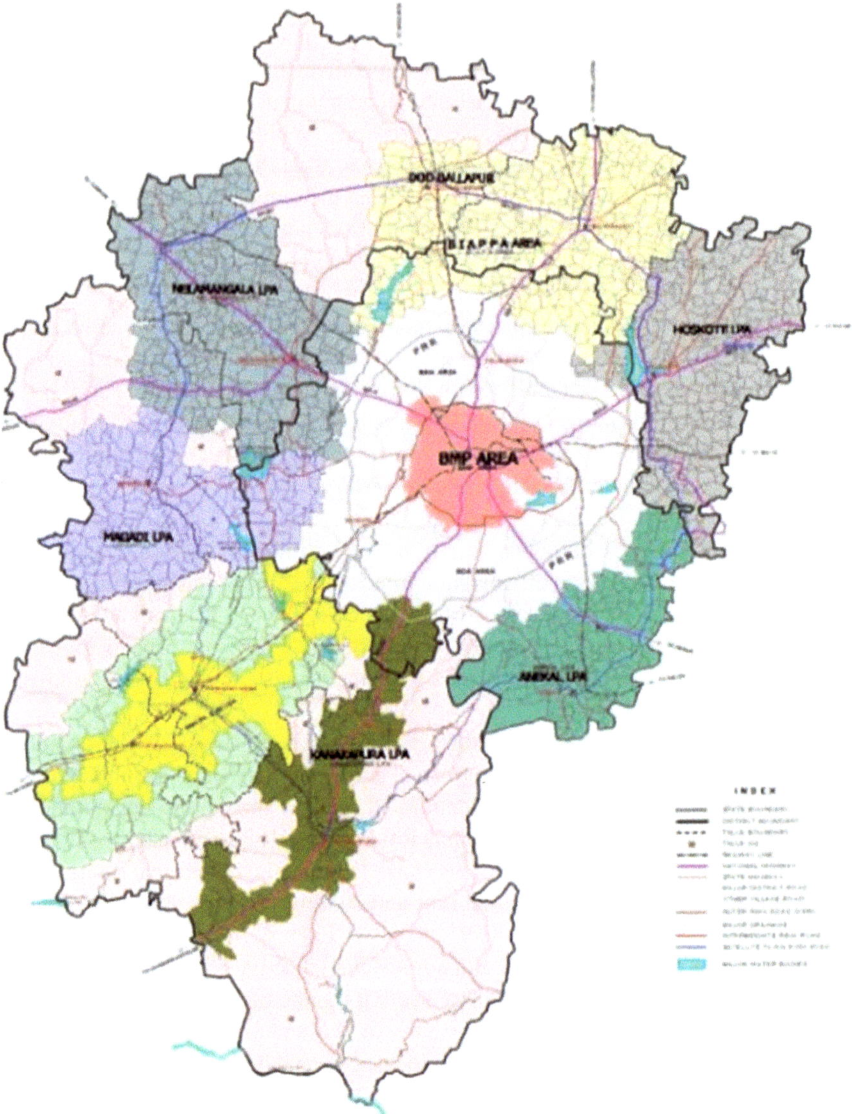

Fig. 1.1 Bangalore metropolitan region with LPAs

other industrial areas, including Vemgal and Narsapura, has witnessed development along the Bengaluru-Chennai belt.

Times of India (Times Property –2015) while analysing the cost of residential accommodations around Hosakote reports the following:

- Connectivity to the airport and Hebbal, coupled with the proximity to established commercial hubs such as Whitefield, ORR and Old Madras Road, have enhanced the potential of the region as an emerging residential market.
- On the Hosakote –Whitefield Road, majority of the existing plot developments are priced between Rs 1000–2000 per sqft, and the apartments slightly off NH-4 on an average are priced at Rs 2200. The apartments on the NH-4 locations towards city cost around Rs 4000 per sqft. Further, a majority of the apartment developments are in the mid-range segment Rs 30–60 lakhs.
- With the continued demand for quality IT-related office spaces, there will be a simultaneous increase in the demand for residential development to meet the needs of the IT sector employees in the area. Hoskote and its surrounding areas can serve as an excellent residential catchment area for both the industrial as well as IT hubs.
- The industrial developments of Narsapur and Malur on and off Old Madras Road (NH-4) have acted as a catalyst to economic growth in the region. The region has already witnessed increasing supply and is expected to witness impressive growth in the backdrop of availability of vacant land parcels and increasing demand. The location holds good potential for growth due to the presence of a number of ongoing gated community developments that will attract more developments to this region.

The main focus was on understanding the various influences at play in Jadigenahalli village council *(Gram Panchayat)* area, located on Hosakote-Malur Road, part of Bengaluru rural district, which could be the situation elsewhere in a semi-arid landscape. The study documented processes and consequences of such influences of urban areas, on the one hand, and on the other, implications of climate change on livelihoods-based activities, followed by coping mechanisms adopted by the local people. Since the study area identified was a semi arid zone, the indicators relevant to the zone were drawn from the various global level studies dealing with the changes happening at the global level with a view to understanding whether there were any linkages in the context of Jadigenahalli village council (*Gram Panchayat*).

The study also addressed health-related interventions made by the government as part of promoting an integrated and holistic health care system through ‘Soukya Foundation’, a private health service provider. The intervention has been made through ‘AYUSH *gram*’ programme, with funds sanctioned to 10 village councils (*Gram panchayats*). Another dimension is linkages between ecology and culture and related health practices that have protected local health traditions and medicinal plants since long. Interestingly, in this village, there is a traditional practice of conserving several medicinal plant species which are considered holy, in addition to using them for treating several health related ailments. Besides health linkages, the study also addressed the linkages between traditional practices that have helped conserve and protect the local ecology. It also captured the negative implications associated with the urban lifestyles in the form of plastic usage, garbage menace etc., affecting the rural ecology and its management. Sanitation is another area the study addresses, documenting positive changes taken place in the village in the context of a near eradication of open defecation with more than 90% of the households having individual toilet facilities.

Research Methodology The study, based on empirical evidences and secondary data, covered five villages from Jadagenahalli village Council (*Gram panchayat).* This *gram panchayat* was selected mainly owing to: its proximity to metropolitan city and is influenced by urbanization; and a high water scarcity witnessed in the surrounding five villages (e.g, dried up rainwater harvesting structures like local tanks, drastic depletion of ground water levels, up to 1200 feet, more than 60% of drilled bore wells have become defunct.

A household survey with structured interviews and focused group discussions was organized at different slots, during December 2013 and April 2014. However, the land-use changes using GIS were done in 2017. The research team made a 3 day pilot field visit to get an overview of the situation besides holding focused group discussions with farmers, traders, women, self-help groups, youth organisations in addition to specific responses about health status, interventions by 'Soukya' (Ayurveda based health services provider with an established local branch), environmental concerns related to water and sanitation, conservation of medicinal plants and their usage. Further, a detailed household structured questionnaire (102 households) and a checklist for focused group discussions (24 FGDs) were used.

Survey Design and Instruments A field survey was conducted at different levels based on the themes and hence contextual. Both structured and semi-structured survey instruments were designed for group-level meetings as well as individual interactions. Primary data was collected from households using a structured questionnaire. The questionnaire design covering socio-economic characteristics, water resources and cropping pattern, food and addicted habits, Soukya activities, preferences and perceptions of healthcare centres, user satisfaction, religious beliefs, medicinal plants and their usage, ecology and cultural aspects was oriented to receiving a systematic public feedback for assessing the implications and perceptions. Based on pilot visits to Jadigenahalli Village Council (*Gram Panchayat*), the study team drafted the questionnaire, both in terms of its contents and the methods of collecting information.

An appropriate sampling design was followed at different levels for selecting respondents for the survey. At the users' level, both structured and semi-structured survey instruments were designed for group-level meetings across types of beneficiaries (like ASHA workers, Self Help group, youth organizations, village leaders, schools and Anganwadis). Similarly, appropriate survey instruments were designed for eliciting information from individual beneficiaries of Ayush Grama Soukya. A sample 100 beneficiaries was drawn from 5 selected villages out of 7 coming under Jadigenahalli Grama Panchayat for a detailed survey (Govindapura, Haraluru, Jadigenhalli, K B Hosahalli, Vadigenahalli). With details of survey and methodology varying across themes covered under the study and has been explained under each chapter. The chapter outline follows a perspective keeping in view the status and interventions.

In continuation with Chap. 1 with an introduction to the peri urban challenges, we present the rest of our chapters with a specific reference to the study area and discuss issues concerning larger contexts.

Chapter 2 focuses on the spatial representation of land use changes, issues of groundwater depletion in the study region and livelihood concerns. The study evidently indicates at unregulated groundwater depletion in terms of a huge drain on household savings, a forced cropping pattern shift from food grains to cash crops for easy cash flows, inevitable out-migration in search of better jobs, in spite of an enormous increase in the local land prices. The chapter also highlights the coping mechanisms adopted by both the landholders and landless across villages, the negligence of common property resources like water storage tanks and their implications for the local ecosystems. During summer, households are forced to buy water for domestic purposes from private tankers; the study analyses the causal factors for this situation besides attempting possible alternative options.

Chapter 3 is an evaluation of the private intervention of health service provider Dr. Mathai's Rural Health Centre (DMRC)- Soukya Foundation under the programme AYUSH Grama; it tries to understand the role of DMRC, Soukya Foundation under the AYUSH programme aimed at reviving the traditional health care systems and also the challenges involved making the development of traditional medicine-based health care a possible solution and how it has fared as a public-private partnership model. Given the bigger challenge of providing rural health care in India with costs and infrastructure challenges being on the rise, traditional medicine is seen as a positive and affordable option for improving the health status of the masses. The chapter also focuses on the activities and status, strengths and challenges encountered in moving towards sustainable approaches in terms of gaining insights into sanitation aspects with a specific reference to toilet facility in terms of coverage and usage and the solid waste management and other sanitation related concerns.

Chapter 4 discusses the cultural practices and traditions that play a significant role in protecting the peri-urban ecology at large under the Jadigenahalli Village Council (*Grama Panchayat*). It also highlights how the emphasis laid on cultural links and ecological importance plays a vital role in maintaining the health status of people and conservation and usage of medicinal plants.

Chapter 5 provides a brief with an overall perspective, highlighting the significance and role of culture and governance in combating climate change.

Annexures

Annexure 1.1: Jadigenahalli – Peri-Urban Fringe of Bengaluru City

Profile of the Study Area

Jadigenahalli Gram Panchayat, Hosakote Taluk, 32.3 kms off the city centre is located in the peri-urban area of growing Bengaluru city forming part of Bengaluru Metropolitan Region. Jadigenahalli Grama Panchayat belongs to Hoskote Taluk, Bangalore Rural District of Karnataka (Fig. 1.2), and is one of the 5 hoblis of Hoskote Taluk (Anugondanahalli, Kasaba, Nandagundi and Sulibele) located at a distance of 13 km from Hoskote Taluk and 37 km from Bangalore city.

Jadigenahalli Gram Panchayat is a small unit of administration in Hosakote, a taluk in Bengaluru Rural District. Agriculture, and horticulture are the primary occupations of people in this area, although ndustrialization in the recent times in Hosakote and places around has offered new employment opportunities for people. Hoskote is located on the eastern side of Bengaluru at the intersection of National Highway (NH) 4 and NH 207 (Fig. 1.2). Hosakote has a close proximity to Whitefield, that is a major IT hub of the Bangalore city, with a slew of IT companies and a large employment base; it has also evolved into a preferable destination for residential and retail development with a surge in the demand for residential units from Whitefield.

Industrial area in Hosakote has nearly 200 industrial units (Master Plan Hosakote) operated by global automotive brands such as VOLVO, Honda etc. Hosakote has a good connectivity to Whitefield (14 KM), Outer Ring Road (ORR) of Bangalore, and lies on the growth corridors of NH- 4 that connects Pune-Bengaluru-Chennai.

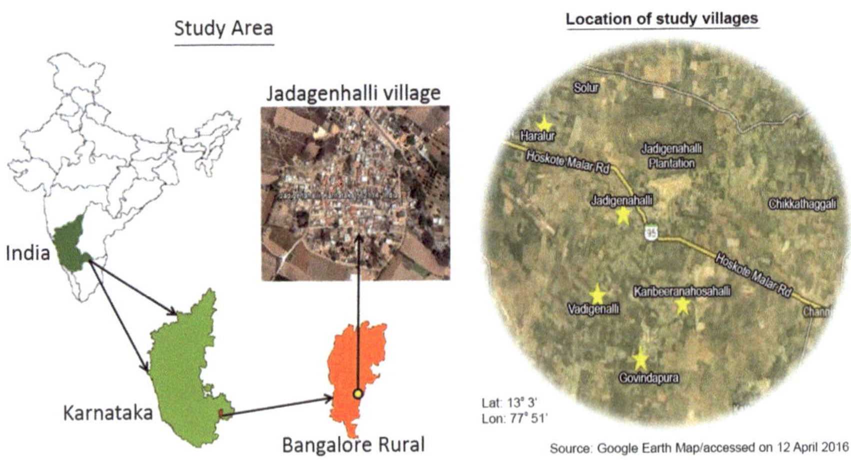

Fig. 1.2 Study area and location of villages

The location has a close proximity to the proposed Peripheral Ring Road expected to reduce the travel time to nearby areas such as Whitefield, Hoodi, Marathahalli etc. upon completion.

There are 7 villages coming under the jurisdiction of Jadigenahalli Grama Panchayat – (1) Jadigenahalli; (2) Vadigenahalli; (3) Karibiranahosahalli; (4) Govindapura; (5) Haraluru; (6) Kolathur; and (7) Kurubaragollahalli. In the present study, we have considered five villages to study the various aspects. Hence, the information and data are restricted to these five villages. Data and information on geographical location, meteorological aspects, socio-economic status, agriculture (Cropping pattern), water sources, sanitation, ecology and culture are discussed in the following sections. (Village details presented in – Annex 1.2) The total geographical area of Jadigenahalli comes to 4155.48 hectares (Ha).

Annexure 1.2: Study Area

The villages coming under Jadigenahalli Gram panchayat are located within a 50 km radius from the growing city of Bengaluru. In 1985, Government of Karnataka constituted Bangalore Metropolitan Region Development Authority (BMRDA) with a statutory limit named Bangalore Metropolitan Region (BMR) with an area of 8005 sq.kms. The region has been further divided into six Local Planning Areas (LPAs) for effecting a better planning and development. Hoskote is one of such LPAs, an important urban center. Hoskote Local Planning Area comprises Hoskote Town and 299 villages under Hoskote Taluk and Jadigenahalli Gram Panchayat being part of the Hoskote LPA is located on Hoskote – Malur Road. The composition and population of the study area are presented in Tables 1.2 and 1.3 (Figs. 1.3, 1.4 and 1.5).

Jadigenahalli is situated at a latitude of 13° 3′ North and a longitude of 77° 51′ east, and is 913 feet above the mean sea level. The temperatures of this regions

Table 1.2 Composition of Jadigenahalli gram panchayat

			Population (no of persons)			
Sl.No	Village name	Area in acres	1981	1991	2001	2011
1	Govindapura	465.94	6	602	810	724
2	Haralur	314.18	109	553	692	832
3	Jadigenahalli	1025.83	1573	1707	1877	1977
4	Jadigenahalli Plantation	452.16				
5	Karibeeranahosahalli	430.7	410	419	532	583
6	Kolathur	1111.56	958	1082	1299	1568
7	Kurubara Gollahalli	526.51	52	33	35	32
8	Vadigehalli	461.12	97	101	120	103
Total		4788	3205	4497	5365	5819

Source: Authors and Census of India, 2011

Table 1.3 Types of housing under Jadigenahalli gram panchayat (%)

House	Type of house			
	Kaccha	Pucca	RCC	Semi Kaccha
Before (1980)	36	3	11	50
After (2013)	10	0	34	56

Source: Authors 2014

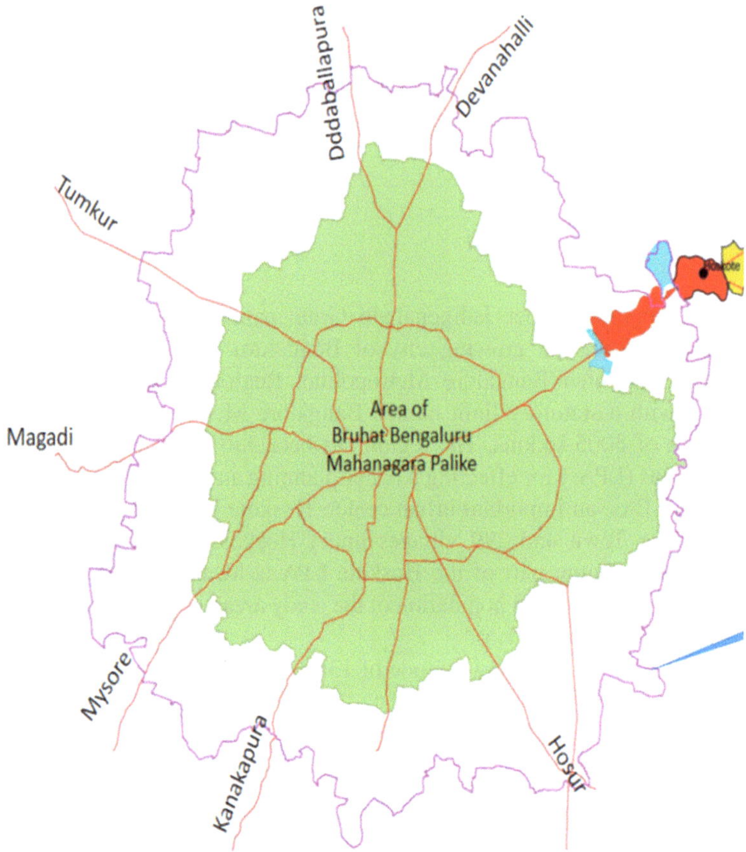

Fig. 1.3 Location of Jadigenahalli gram panchayat

range from 32–36° Celsius during the summer season (IMD) and in the rainy season from 28–30° Celsius and from 27–28° Celsius during the winter season. The normal average annual rainfall pattern of Jadigenahalli Grama Panchayat amounts to 750.4 mm. However, the rainfall pattern, on an average, over the period 1989–2013 shows normal rainfall (750.99 mm). The figure below indicates that rainfall recorded was above normal for 1991, 1993, 2005 and 2008, with a declining trend observed for 2012 and 2013 (Annex 1.3).

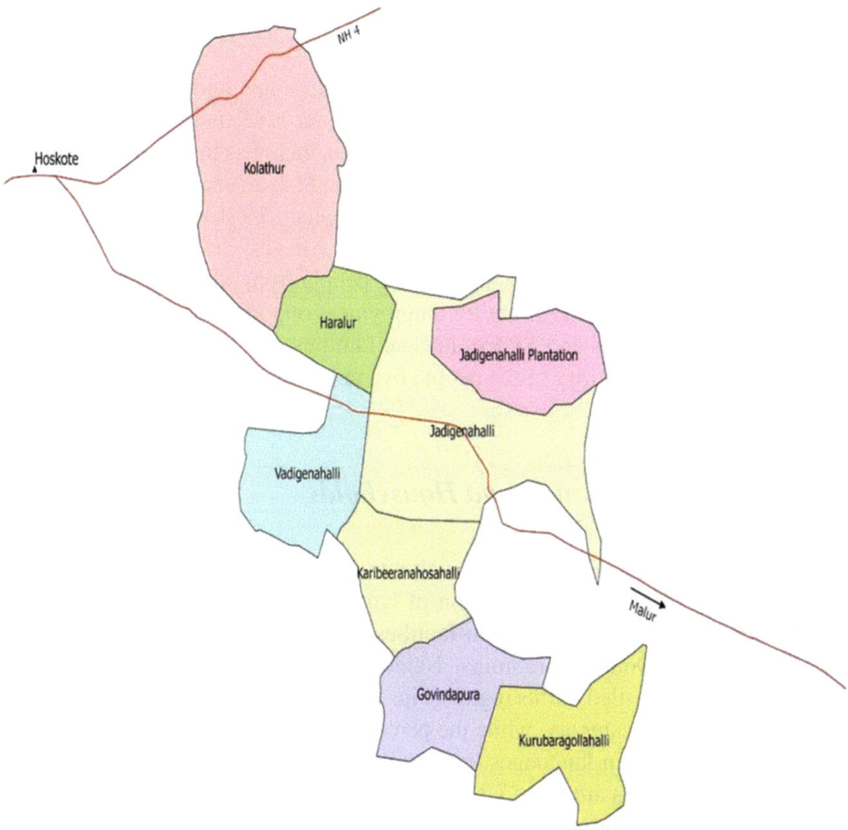

Fig. 1.4 Villages Surveyed under Jadigenahalli gram panchayat. (Source: Google earth map, Scale 1:2 km)

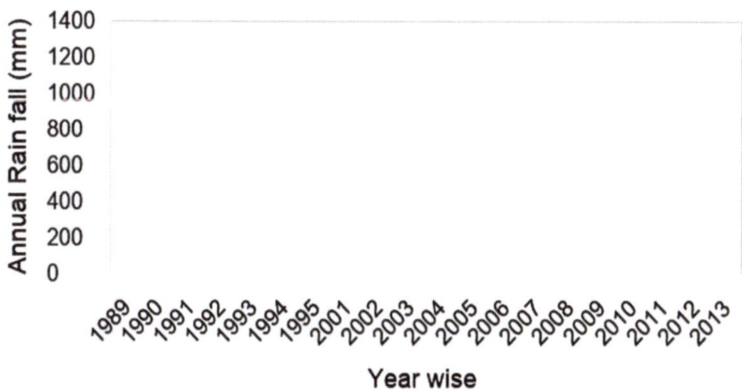

Fig. 1.5 Annual rainfall

Annexure 1.3: Socio-Economic Status

A majority of the people are dependent on agriculture and agricultural labour for their livelihoods. Dependency on private employment is second in order (factories and brick making units). Over the past two decades, there has been a drastic change observed in the development activities in and around Jadigenahalli GP. Since villages are located nearer (10 Km) to the urban towns – Hoskote and Bangalore – there is a lot of urban influence observed on the villages. Several industries also have come up over the last 20 years, attracting people from other states like Orissa, West Bengal, Rajasthan and Assam for employment opportunities. Jadigenhalli GP has 47% of Joint families and 53% of nuclear families. In Vadigehalli, Jadigenhalli and Govindhpura, a majority of the people live in joint families.

Annexure 1.4: Population and Households

There is a marginal increase in the growth rate of population, while the number of households has increased with division of land over generations. The people are mostly Hindus, with a few Christian families residing in these villages. Gowdas, Thigalas, Bhovis, Dobbis, Brahmins, Naidus and Kurubas constitute the major community groups settled in this region (Fig. 1.6). A large proportion of the population belongs to OBC category, while the percentage of SC and ST is comparatively less. Commonly spoken languages include Kannada, Telugu and Tamil. RCC and Semi Kaccha buildings are more visible as compared to other structures with electricity as a common lighting source. For cooking, 75–80% use LPG and the rest, fire wood and Kerosene.

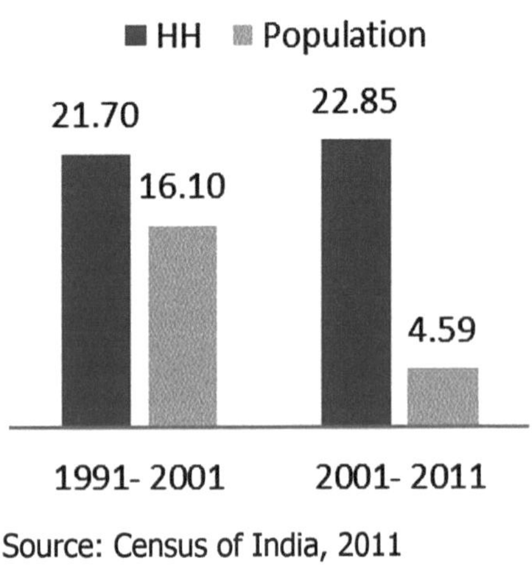

Fig. 1.6 Growth rates of Jadigenahalli in terms of HHs and population at two different points of time (%)

Annexure 1.5: Literacy Rate

Literacy rates have gradually increased over a period of two decades. The impact of Karnataka's target of providing primary education by 2007 and higher primary education by 2010 seems to have led to a positive trend (Fig. 1.7). Increased investments in terms of time, efforts and resources are seen as the main cause for an increase in the literacy rates in Karnataka (*Human Resource Karnataka, ISEC*). Besides, mid-day meals programme being implemented since 1995 by the Central Government as part of encouraging enrolments, retention and attendance also seems to have contributed to literacy improvements in the state, including these study villages. Access to towns and Bangalore through good transportation facilities has made a positive influence on the literacy rates among the populace in the villages, reaching up to 71% literacy by 2011.

Annexure 1.6: Gender Status

An interesting point to be noted in respect of these villages is that the female population is found higher than male population for 1991 and 2001, but for 2011, this situation seems slightly different due to family planning programmes initiated by the Government. The literacy rates of the villages for 1991 and 2011 have also changed, indicating that education and urbanisation have influenced them in preferring one or two children per family which was not the case earlier. During earlier times, the general view of people was that a family would be complete only with the birth of a male child and hence, irrespective of the number of girl children born, they would continue their efforts until the birth of a male child. This could be the reason behind an increase in the number of females.

Fig. 1.7 Literacy rate

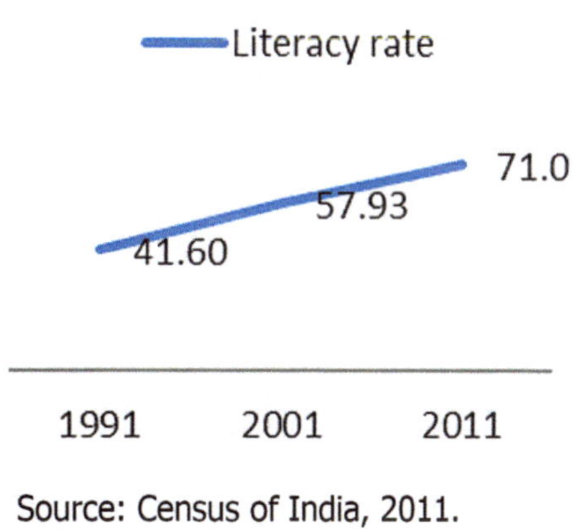

Annexure 1.7: Domestic and Drinking Water Sources

People were dependent on open wells as a major source of drinking and domestic water until 1991 (Haraluru, Vadigenahalli, Karibeeranahosahalli and Govindapura) with Jadigenahalli having a piped water connection sourced from a bore well. After 2001, the sources changed to tube wells and tanks in Jadigenahalli and hand pumps in Vadigehalli and Govindapura (*See Annexure*). Currently, the common source of water (for drinking and domestic usage) across all the villages is piped water supply sourced from bore wells under the PWS scheme of the Grama Panchayat. All the five study villages come under the semi-arid region of Karnataka with ground water as the only source for agriculture, domestic use and livestock activities. In the absence of other sources of water, there is a considerable pressure observed on ground water resource. Our discussions with the respondents reveal that, in all the study villages, tanks have dried up since many years. For drinking water, about 80% of the HHs are dependent on piped water supply by the Government through *panchayat,* while about 54% of the HHs are dependent on the same source for livestock. The supply of water for domestic use is not sufficient with many disputes observed among the communities over sharing of water. For agriculture, 56% of the HHs have own bore-wells for irrigation.

References

Bahadur Aditya, 'Peri Urban: Is the unwanted child of 'Rural and Urban'?, National Conference on Peri-Urban Agriculture and Ecosystems, The Multifaceted contribution to Urban Resilience', 22–23 Feb 2016, Organized by Gorakhopur Environmental Action Group in collaboration with the National Institute of Urban Affairs, Government of India and the RUAF foundation at India International Centre, New Delhi (2016)

M. Amit, S. Wajih, B. Singh, Wheezing Ecosystem, Livelihood Services, Climate Change Resilience in Uttar Pradesh, Working Paper Series 18, Iied, The Rockfeller Foundation (2015)

A.L. Clark, Environmental Challenges to Urban Planning: Fringe Areas, Ecological Footprints and Climate Change, paper presented at workshop 'Key Challenges in the Process of Urbanization in Ho Chi Minh City-Governance, Socio-Economic and Environmental Issues, 16–18th September 2008, Ho Chi Minh City, Vietnam, (2009)

R.N Datta, Talk during round table on 'Peri Urban Areas around Indian Cities' on July 1st 2014 at IIT Bombay (2014). http://cuse.iitb.ac.in/development-of-peri-urban-areas-around-indian-cities/

H. De Zeeuw, R. Van Veenhuizen, M. Dubbeling. The role of Urban Agriculture in building Resilient Cities., 153163 FAO (2007) Coping with water scarcity challenge of the twenty-first century. UN Water (2011). See www.fao.org/nr/water/docs /escarcity.pdf

I. Douglas, Environmental change in Peri-urban areas and human and ecosystem health. Geogr. Compass **2**(4), 1095–1137 (2008)

FAO, Food Agriculture and Cities, Challenges of Food and Nutrition security, agriculture and ecosystem management in an urbanizaing world in FAO Food for the Cities multi-disciplinary initiative, position paper, 2011

S. Lwasa, M. Tenywa, G. J. Majaliwa Mwanjalolo, G. Prain, & H. Sengendo, Enhancing adaptation of poor urban dwellers to the effects of climate variability and change. 3 (Vol. 6, p. 332002). Presented at the IOP Conference Series: Earth and Environmental Science (2009)

F. Marshall; L. Waldman; H. MacGregor; L. Mehta and P. Randhawa (2009) On the Edge of Sustainability: Perspectives on Peri-Urban Dynamics. STEPS Working Paper 35. See www.ids.ac.uk/publication/on-the-edge-of-sustainabilityperspectives- on-peri-urban-dynamics

D. Mkwambisi Urban Agriculture and Food Security in Lilongwe and Blantyre, Malawi, in Redwood, M. (ed), *Agriculture in Urban Planning: Generating Livelihoods and Food Security,* London: Earthscan Satterthwaite, D., Huq, S., Pelling, M., Reid, H. and Lankao, (2009)

C. Mutonodzo, The social and economic implications of urban agriculture on food security in Harare, Zimbabwe, in *Agriculture in Urban Planning: Generating Livelihoods and Food Security*, ed. by M. Redwood, (Earthscan, London, UK, 2009)

V. Narain (2007), Between rural and urban: Policy in practice for the peri-urban interface. Paper presented at the Second Annual Conference on Public Policy and Management, April 12–14, 2007. Bangalore: Indian Institute of Management. p. 24

V. Narain Peri-urban water security in a context of urbanization and climate change: A review of concepts and relationships. Peri Urban Water Security Discussion Paper Series 1, SaciWATERs. (2010) See http://saciwaters.org/periurban/idrcperiurbanreport.pdf

National Bank for Agriculture and Rural Development (2018), NABARD all India Rural Financial Inclusion Survey 2016–17 (NAFIS). Department of Economic Analysis and Research, NABARD, Mumbai, India

K. Pothukuchi, J.L. Kaufman, Placing the food system on the urban agenda: The role of municipal institutions in food systems planning. Agric. Hum. Values **16**(2), 213224 (1999)

A. Prakash, The periurban water security problématique: a case study of Hyderabad in Southern India. Peri Urban Water Security Discussion Paper Series 2010/ 4, SaciWATERs (2012). See www.saciwaters.org/ working_papers.html

K. Ratoola, Talk during round table on 'Peri Urban Areas around Indian Cities' on July 1st 2014 at IIT Bombay. http://cuse.iitb.ac.in/development-of-peri-urban-areas-around-indian-cities/

D. Roberts, R. Boon, N. Diederichs, E. Douwes, N. Govender, A. McInnes, C. McLean, et al., Exploring ecosystem-based adaptation in Durban, South Africa: "Learning-by-doing" at the local government coal face. Environ. Urban. **24**, 167–195 (2011). https://doi.org/10.1177/0956247811431412

Srivastava Ravi S.K. Sasikumar (2003), 'An overview of migration in India, its impacts and key issues, paper prepared for Regional Conference on 'Migration Development Pro poor Choices in Asia, on 22–23 June 2003, Dhaka, Bangladesh

R. Usha, 'Peri Urban: Is the Unwanted Child of 'Rural and Urban'?, National Conference on Peri-Urban Agriculture and Ecosystems, the Multifaceted Contribution to Urban Resilience', 22–23 Feb 2016, Organized by Gorakhopur Environmental Action Group in Collaboration with the National Institute of Urban Affairs, Government of India and the RUAF foundation at India International Centre, New Delhi (2016)

Websites

https://www.environment.gov.za/sites/default/files/docs/ltasphase report_adaptation human settlements.pdf

https://books.google.co.in/books/peri urban areas and climate change & source one page

http://www.nwo.nl/en/research-andresults/programmes/Conflict+and+Cooperation+over+Natural+Resources+in+Developing+Countries+%28CoCooN%29/Project+Dr+Roth

https://www.rockefellerfoundation.org/app/uploads/Degredation-and-Loss-of-Peri-Urban-Systems.pdf

http://start.org/download/gec11/lwasa-final-report.pdf

http://www.plurel.net/images/PLURELsynthesis report SUMMARY.pdf

http://web.ntpu.edu.tw/~shuli/PDF/PUGEC/02Douglas.pdf

http://web.mit.edu/urbanupgrading/urbanenvironment/issues/rural-urban-linkages.html

https://www.un.org/development/desa/en/news/population/2018-revision-of-world-urbanization-prospects.html

growing population have been exerting a considerable stress on the surrounding landscape/ environment. Land use changes are mostly due to intensive human activities, which can be observed using current and archived remotely sensed data.

As discussed earlier, an exponential increase in the population growth in Bengaluru region has resulted in the increasing demands for land in the peri-urban areas of Bengaluru. Hoskote area on the outskirts of Bengaluru City has been one of regions experiencing change in land use from agriculture to non-agricultural use resulting in a decrease in vegetation and water bodies and thereby to depleting groundwater levels. A study was carried out to understand the present overlay of land use and land cover in Hoskote Taluk, Kasaba Hobli (Srimani and Nanditha Prasad 2013) focussing on the effectiveness of the satellite data for land use/land cover using multi-temporal images (Seasonal data). However, long-term changes in the landscape have not been attempted for the important peri-urban areas. Also, a few studies have been conducted with regard to groundwater which, in the recent years, has become the only source for varied requirements. The current chapter is based on a study (conducted in 2017) an attempt to analyse the land use changes and pressure on groundwater (in terms of the number of structures and their spatial distribution) under Jadigenehalli Gram Panchayat area. This would support the findings from the spatial data and also help have a better understanding of the field realities in terms of the implications of groundwater depletion and related issues. The chapter is also supported by earlier field level study (conducted in 2014) involving interviews with farmers living in the villages coming under Jadigenahalli *Gram Panchayat*.

2.2 Methodology

Input data/materials: (a) Survey of India Toposheet (1:50000) of 1973 edition; (b) Taluk Map of Hoskote Taluk (Survey Settlement and Land Records of Karnataka); (c) Multispectral satellite images available on Google Earth (2003, 2009, 2013 and 2017); (d) Field enumeration of wells.

Taluk map of Hosakote has been used to identify the villages of Jadigenahalli Gram Panchayat. The boundary of Jadigenahalli so derived has been registered with the topo sheet and further with satellite images for a proper land use analysis in a GIS environment.

2.2.1 Land Use Classes

A standard land use classification scheme adopted by NRSC (2014) for nationwide mapping has been followed. Land use derived from the topo sheet (1973) has been kept as reference base information. Although more resolved classes of land use

Table 2.1 Land use/land cover schema adopted for Jadigenahalli

Sl. no	Main land use class	Details included in the class
1	Built up	Scattered buildings
		Settlements
		Poultry farm
		Industries
2	Brick kiln	Brick kiln
3	New layout	New layout
4	Forest land	Forest plantation
5	Agriculture land	Crop land
		Agricultural plantation/garden
6	Water bodies	Tanks
7	Rocky area	Rocky land
		Stone quarries/crushers
8	Commercial plantation	Eucalyptus plantation
9	Others	Area covered by stream course, scrubland and roads

could be derived from the interpretation of satellite images, for comparison purpose only a few have been retained. The land use features derived from images related to different years have been finalised based on the ground validation (ground truth). The land use information derived has been finally grouped under different classes as shown in Table 2.1.

A standard reference signature (image characteristics) has been generated using images of different years so as to maintain consistency in the interpretation of different land use classes. The area of Tanks and forests (forest plantation) has been taken from the toposheet. On screen digitisation of different land use classes have been made and the data organised in a GIS environment. QGIS has been adopted to integrate the spatial information for different years and the area calculation done for a further change analysis.

2.2.2 *Description of Land Use Classes*

The built up land constitutes area used for general dwelling purposes and other uses covered with buildings is classified as Built up. In respect of Jadigenahalli, this includes village settlements, scattered buildings, poultry farms and industrial sheds/buildings. A few brick kilns have been observed and as they occupy a considerable area, a separate class has been derived and estimated. A substantial area has been converted into residential layouts in the recent years. Currently most of them have not been completely occupied. Under the Jadigenahalli *Gram Panchayat*, a certain area declared as forest land is currently occupied by eucalyptus plantations.

Since there is a clear boundary demarcation (in the 1973 toposheet), it is considered as forest land. A huge area observed near Jadigenahalli is called Jadigenahalli Plantation in the revenue records. Under the Agriculture Land, two classes have been distinguished under this category. First one is normal crop land being cultivated in any one of the growing seasons with rainfall as a source of water and the second is agriculture plantation/garden, which is essentially a perennial type supported by groundwater. Water Bodies (Tanks), they have been grouped under this category. Although currently these tanks do not have appreciable water spread, the area taken from the toposheet of 1973 has been used as the reference point and based on the image characteristics; area of tanks has been derived. The image characteristics have been used for identifying areas that do not have any vegetative cover, exposing a rocky surface. At places, rock quarries have been created and these two have been clubbed as one class. Commercial Plantations have replaced most of the agricultural land. The images show distinctly these features and have been interpreted. The size and shape and tone are the key elements used for the interpretation of such plantations. There are others like, the areas occupied by stream courses, roads and scrublands (not used for agriculture) have been clubbed under this category.

2.2.3 *Well Inventory*

The Jadignehalli (composite of all the villages) map was converted in to georeferenced tile map and was loaded on to Smartphone. An android application was developed to capture the coordinates of all the wells (both bore wells and open wells) and a few attributes. Ground validation staff carried the smart phone, traversing across all the villages and capturing a few attributes. Following were the attributes that could be captured:

1. Location of the structure
2. Depth of the structure (in case of an open well, it was measured and for bore well, the information was gathered from farmers who were available at the site)
3. Status of the structure – defunct or functional
4. Historical aspects – Year of drilling/ year of becoming defunct/ area irrigated: In most of the cases, the response was not available as all the farmers were not able to remember exactly the year of drilling and few were not available during the inventory
5. Yield – In most of the cases, only qualitative information was available and hence the information was used only for comments

The data was integrated with the GIS data of Jadigenahalli and the location of each well plotted. The spatial distribution map of wells with bore well depth variations was derived and plotted. Also a correlation of location of wells with land use category was also carried out.

2.3 Results

The land use data as well as the groundwater abstraction structures were organised on an open source GIS platform i.e., QGIS. Different layers for every land use classes were created and the area statistics generated. The results are presented in the following sections. The spatial data derived related to land use for different years is provided in Tables 2.2 and 2.3 and for 2017 (village wise) in 2.4. (see Annex 2.1, 2.2, 2.3, 2.4, 2.5 land use for each village and for the entire village council (*Gram Panchayat*)) (Table 2.4). The maps generated based on the interpretation of toposheet and satellite images are also presented in Figs. 2.1, 2.2, 2.3, 2.4, 2.5. Field photographs of different land use classes are shown in Plate 2.1, 2.2, 2.3, 2.4, 2.5, 2.6, 2.7, 2.8.

2.3.1 *Land Use Changes*

Analysis of the land use over the years is presented in the following sections

Built-up Land As the population has increased, settlement area has also increased and hence, the area under built-up category has increased by almost four times as

Table 2.2 Land use/land cover over the years in Jadigenahalli area in acres

Main class	Sub class	1973	2003	2009	2013	2017
Built up	Scattered buildings		11.96	9.8	21.71	21.05
	Settlements	51.06	84.31	104.13	102.57	115.7
	Poultry farm		24.54	49.51	35.94	36.55
	Industries		4.48	25	28	22.23
	Sub-total of built up land	51.6	125.28	188.45	188.21	195.53
Brick kiln	Brick kiln		1.17	10.39	13.22	13.19
New layout	New layout		21.99	25.59	74.61	109.26
Forest land	Forest plantation	722.4	722.3	722.52	722.79	722.22
Agriculture land	Crop land	3627.22	1465.73	1404.66	1314.47	1225.58
	Agri. plantation		313.06	259.56	280.45	279.87
	Sub-total of agriculture land	3627.22	1778.79	1664.22	1594.92	1505.45
Water bodies	Tanks	232.22	199	178.58	181.24	177.56
Rocky area	Rocky land	4.14	11.62	12.11	11.9	11.94
	Stone quarries/crushers		2.29	5.26	6.72	7.09
	Sub-total of rocky area	4.14	13.91	17.37	18.62	19.03
Commercial plantation	Eucalyptus plantation	123	1543.85	1693.19	1677.98	1646.13
Others	Area covered by stream, road, scrub land	27.99	381.67	287.47	316.42	399.63
Total		4788	4788	4788	4788	4788

Table 2.3 Land use pattern under Jadigenahalli gram panchayat: 1973–2017

Land use class	Year wise statistics (Area in acres)									
	1973		2003		2009		2013		2017	
	Area	%	Area	%	Area	%	Area	%	Area	%
Built-up	51.06	1.07	125.28	2.62	188.45	3.94	188.21	3.93	195.54	4.08
Brick kiln			1.17	0.02	10.39	0.22	13.22	0.28	13.19	0.28
New layout			21.99	0.46	25.59	0.53	74.61	1.56	109.26	2.28
Forest land	722.4	15.09	722.3	15.09	722.52	15.09	722.79	15.10	722.75	15.10
Agriculture land	3627.22	75.76	1778.79	37.15	1664.22	34.76	1594.92	33.31	1505.45	31.44
Water bodies	232.22	4.85	199	4.16	178.58	3.73	181.24	3.79	177.56	3.71
Rocky area	4.14	0.09	13.91	0.29	17.37	0.36	18.62	0.39	19.03	0.40
Commercial plantation	123	2.57	1543.85	32.24	1693.19	35.36	1677.98	35.05	1645.6	34.37
Others	27.99	0.58	381.67	7.97	287.47	6.00	316.42	6.61	399.63	8.35
Total area	4788		4788		4788		4788		4788	

Table 2.4 Current land use pattern (2017) under Jadigenahalli GP – % distribution of classes across villages

Land use classes	Govindapura		Haralur		Karibeerana hosahalli		Kolathuru		Kurubara Gollahallii		Vadigehalli		Jadigenahalli		J. Plantation	Total of Jadigenahalli GP	
	Area in acres	%	Area in acres	%	Area in acres	%	Area in acres	%	Area in acres	%	Area in acres	%	Area in acres	%	Area in acres	Area in acres	%
Built-up	22.6	4.85	19.78	6.30	3.78	0.88	69.08	6.21	6.29	1.19	4.9	1.06	69.1	6.74		195.54	4.08
Brick kiln	1.33	0.29	0.95	0.30	2.24	0.52	6.16	0.55	2.51	0.48						13.19	0.28
New layout			1.77	0.56			52.79	4.75	2.97	0.56	4.88	1.06	46.84	4.57		109.26	2.28
Forest (land) plantation							38.4	3.45			231.7	50.25			452.16	722.26	15.08
Agriculture land	91.82	19.71	138.54	44.10	187.64	43.57	386.88	34.81	185.05	35.15	115.59	25.07	399.89	38.98		1505.4	31.44
Water bodies	23.48	5.04	13.95	4.44	12.61	2.93	69.25	6.23	0.51	0.10	0.76	0.16	56.99	5.56		177.56	3.71
Rocky area	0.11	0.02	1.07	0.34			7.09	0.64	10.76	2.04						19.03	0.40
Commercial plantation	275.35	59.10	125.64	39.99	192.14	44.61	397.92	35.80	271.72	51.61	29.84	6.47	353.52	34.46		1646.13	34.38
Others	51.25	11.00	12.49	3.98	32.28	7.49	83.98	7.56	46.7	8.87	73.45	15.93	99.48	9.70		399.63	8.35
Total area	465.94		314.18		430.7		1111.56		526.51		461.12		1025.8		452.16	4788	

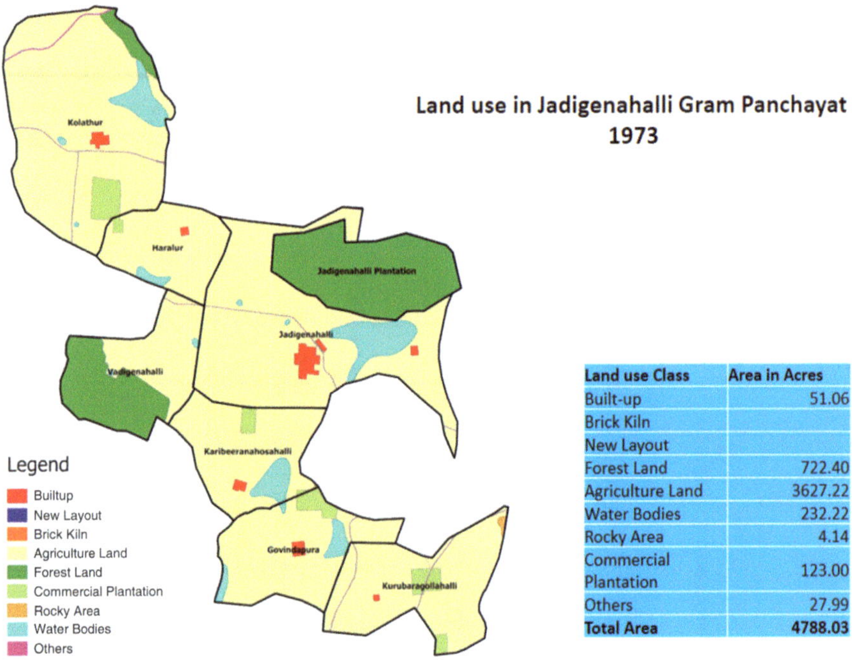

Land use Class	Area in Acres
Built-up	51.06
Brick Kiln	
New Layout	
Forest Land	722.40
Agriculture Land	3627.22
Water Bodies	232.22
Rocky Area	4.14
Commercial Plantation	123.00
Others	27.99
Total Area	**4788.03**

Fig. 2.1 Land use in Jadigenahalli during 1973

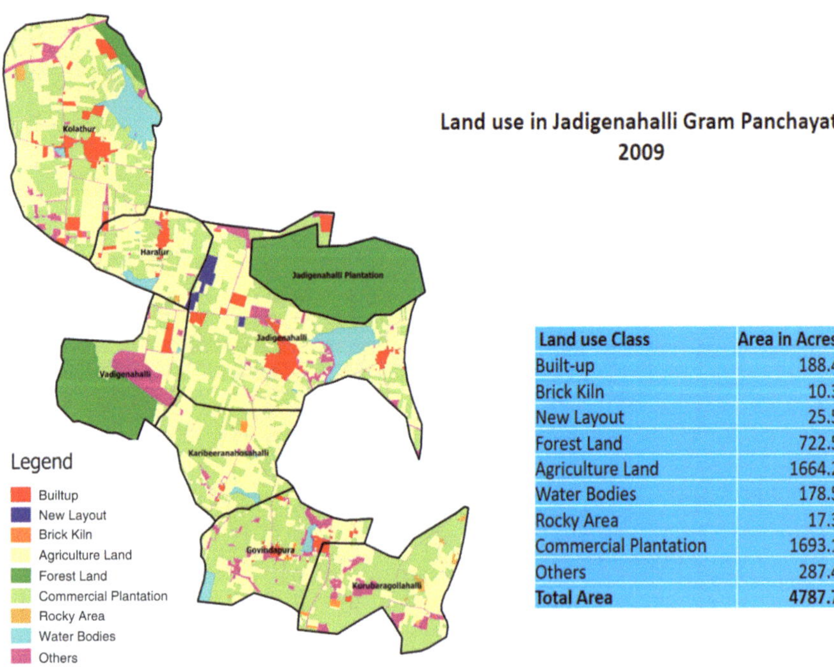

Land use Class	Area in Acres
Built-up	188.45
Brick Kiln	10.39
New Layout	25.59
Forest Land	722.52
Agriculture Land	1664.22
Water Bodies	178.58
Rocky Area	17.37
Commercial Plantation	1693.19
Others	287.47
Total Area	**4787.78**

Fig. 2.2 Land use pattern under Jadigenahalli – 2003

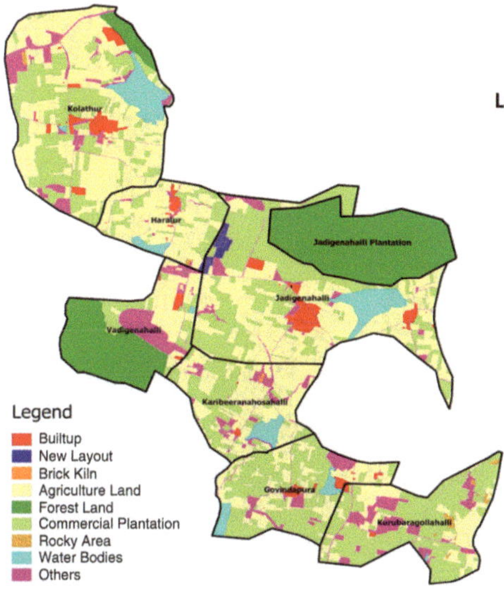

Land use in Jadigenahalli Gram Panchayat 2003

Land use Class	Area in Acres
Builtup	125.28
Brick Kiln	1.17
New Layout	21.99
Forest Land	722.30
Agriculture Land	1778.79
Water Bodies	199.00
Rocky Area	13.91
Commercial Plantation	1543.85
Others	381.67
Total Area	**4787.96**

Fig. 2.3 Land use pattern under Jadigenahalli – 2009

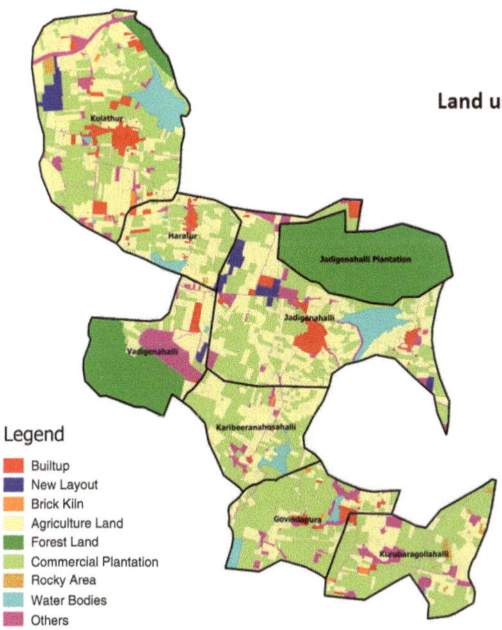

Land use in Jadigenahalli Gram Panchayat 2013

Land use Class	Area in Acres
Builtup	188.21
Brick Kiln	13.22
New Layout	74.61
Forest Land	722.79
Agriculture Land	1594.92
Water Bodies	181.24
Rocky Area	18.62
Commercial Plantation	1677.98
Others	316.42
Total Area	**4788.02**

Fig. 2.4 Land use pattern under Jadigenahalli – 2013

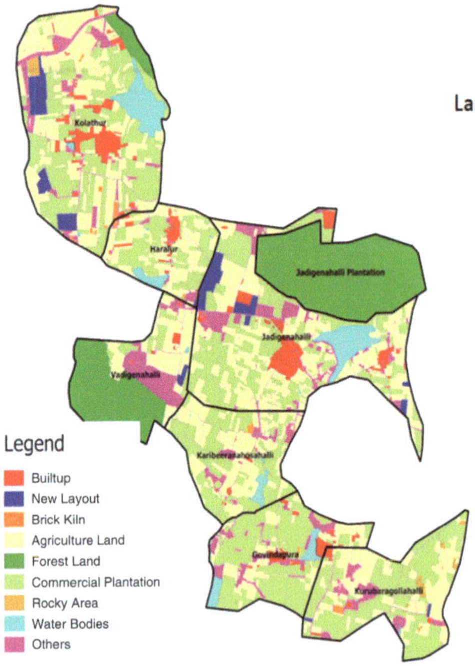

Land use in Jadigenahalli Gram Panchayat 2017

Land use Class	Area in Acres
Built-up	195.54
Brick Kiln	13.19
New Layout	109.26
Forest Land	722.75
Agriculture Land	1505.45
Water Bodies	177.56
Rocky Area	19.03
Commercial Plantation	1645.60
Others	399.63
Total Area	4788.00

Fig. 2.5 Land use pattern under Jadigenahalli – 2017

Plate 2.1 Brick kiln near Govindapura

Plate 2.2 Unoccupied residential layout near Kolathur

Plate 2.3 Eucalyptus plantation near Jadigenahalli (Forest Plantation)

compared to 1973. However, a few industrial establishments have also come up along the road from Hosakote to Malur. As the city of Bengaluru started expanding, the NH leading to Kolar has been the loci of ribbon development and the same has reached Hosakote. The road leading to Malur also has been witnessing such developments. One can observe that the villages like Kolathur, Haralur, Jadigenahalli located along the Hosakote and Malur road have witnessed increased built-up lands.

Plate 2.4 Kharif crop (Ragi) near Vadigehalli

Plate 2.5 Grape garden near Jadigenahalli

As development activity has increased over the years, so has the need for building materials. The result is that around agricultural lands in the vicinity of villages and places near to tanks, brick kilns have come up. Also, the increased demand for

Plate 2.6 Rocky area near Haralur

Plate 2.7 Eucalyptus plantation in agriculture land near Govindapura

residential areas at affordable cost has spread all around Bengaluru City. With transportation facility being improved, the demand for housing also has increased in the surrounding regions. An indication of such pressure is the development of new residential layouts. Real estate entrepreneurs have developed a substantial area as residential layouts under Jadigenahalli *Gram Panchayat* on the Hosakote to Malur road,

Plate 2.8 Scrubland near Jadigenahalli

though most of them have not been occupied. The new layouts have come up in Kolathur, Haralur, Jadigenahalli, Vadigehalli and to a small extent in Kurubara Gollahalli. It can be observed with the Hosakote – Malur Road passing through these villages, Haraluru and Kolathur has also witnessed the influence of National Highway leading to Kolar.

Forest Land Excepting the notified forest (as per the toposheet of 1973), there appears to be no increment to the forest area. The Jadigenahalli Plantation and other few such forest lands have remained under eucalyptus plantations.

Agriculture Land The statistics for different years on agricultural land indicates clearly that the focus on general crop cultivation has come to be replaced with eucalyptus plantations. The agricultural land (crop land and gardens/plantations), which was 3622.77 acres has declined to 1505.45 acres, almost by half over the years. The garden lands like grape and other agricultural plantations must have come up in between 1973 and 2003. However, it can be observed that these also have declined over time.

Agricultural Garden Agricultural gardens could not be interpreted based on the toposheet of 1973. However, the satellite images have clearly indicated the presence of such features for the rest of the years. The area covered by agricultural gardens varies from less than 1% in Govindapura (2009) to 10.6% in Karibeeranahosahalli (2009). Kolathur village has more area under this category (more than 10% of the village area). Although a slight variation in the average percentage of area covered by agricultural gardens over the years has been observed, the trend indicates a slow reduction in the area.

Water Bodies Most of the tanks in the area do not have water. It is only during rains that they are able to store local run-offs temporarily. During the field study, we observed most of the tanks being silted up, and as such, they are not providing their services to the local communities.

Rocky Area The area under this class has been slowly increasing. At places mostly due to greater soil erosion, the rocky surfaces have been exposed with their being quarried for construction material at many places.

Commercial Plantation During 1973, a very small area was under eucalyptus plantations as shown in the toposheet. Later on, there has been (during 1973–2003 i.e., almost 30 years) a gradual increase observed in the area under eucalyptus plantations as in other parts of the state. In view of a good demand from rayon industries, people are being attracted to this less labour intensive and maintenance free cultivation without understanding the implications for local climate and water resources. As of 2017, almost one third of Jadigenahalli GP area has been brought under eucalyptus. Almost all the villages have more than 30% of the area covered by eucalyptus with Govindapura recording the highest coverage at 59.1%, while Vadigehalli has only 6.5% of the area under eucalyptus.

During 1973, the area covered under eucalyptus was very less (6.4% in Kurubara Gollahalli) but over the years, it has reached to almost 1/3 of the area. The village Vadigehalli has shown a decreasing area under eucalyptus.

Others The area covered by streams and scrubland and roads has been grouped under this category. No additional streams or canals are seen in the area and roads also though increased have been limited to connecting villages. The increase in scrublands over time is a direct indication of agricultural land being left vacant for long. Although ownership of such lands could not be ascertained, it is evident that cultivation practice has given way for non-agricultural professions.

2.3.2 Groundwater Over Exploitation

The severity of an impending water crisis at the global level, in general and the Asian region, in particular, is such that it can have serious implications for food security, sustainability of agriculture and the livelihoods of millions of human beings. Water resources have been experiencing an intense and sustained demand pressure from a range of direct and indirect socio-economic driving forces. Arguably people have been treating water almost as a free resource, despite the fact that competition for water accessing is intensifying. While globally, the absolute physical scarcity of water is at best a long-term concern, the current management of water resources has been found wanting in terms of being inefficient, inequitable and environmentally damaging. While the agricultural sector is often cited as the principal 'user' of water, domestic, municipal and industrial uses of water are increasing, though there is now a more widespread recognition of the important environmental services provided by water resources. Thus, the management of water has become an economic, social and political issue.

Groundwater is a critical resource in India, accounting for over 65% of irrigation water and 85% of drinking water supplies (World Bank 2010). However, based on current trends, it is estimated that 60% of groundwater sources will be in a critical state of degradation within the next 20 years (World Bank 2010). The case of India is distinct as it dominates with the highest share of 19.5% in the fresh water withdrawn in the world though its share in agriculture (86.5%) being fairly less than a couple of other countries (World Development Report 2011). In India, groundwater depletion is an issue for concern because it is expensive and increases the vulnerability of cost to the society. The costs of groundwater depletion refer to capital losses resulting from bore well infrastructure being ineffective due to lower groundwater levels, new investments required to deepen the existing bore wells, and new bore wells, higher energy costs involved in groundwater withdrawal and the indirect costs of lower net incomes from reduced irrigated areas. Irrigated agriculture is a major driver of increased water scarcity in view of its high consumption of water resources.

In India, a growing ground water scarcity due to the depletion of aquifers has led to negative externalities- frequent bore-well failures (cumulative well interference), falling water tables, etc. India's average annual rainfall is extremely abundant by global standards, yet much of this rain falls in relatively brief deluges during the monsoon with significant variations across different regions. A combination of these climatic conditions and a range of man-made pressures have driven India's farmers, households, and industry to increasingly depend on groundwater rather than surface water sources like rivers and lakes. But this dependence is leading to a rapid and worrisome deterioration in the nation's groundwater resources.

In the absence of surface water for irrigation, ground water is under pressure as there is a race for exploitation of the resource by those who can afford to drill bore wells and tap water from deeper aquifers. There are also concerns emerging from the unregulated exploitation of ground water resources across states especially, Rajasthan, Gujarat, Punjab, Maharashtra and parts of Karnataka. Under Indian law, the ownership of land carries with it the ownership of ground water under it, subject to regulation and control by the state. Since the right over ground water use is also attached to land property, there are difficulties in regulating the way in which individuals extract the ground water resource for their use.

In rural areas, electricity subsidies allowing farmers to pump out groundwater cheaply have become entrenched in the political landscape (Shah T 2011). They are likely to become even more so as energy requirements increase, forcing farmers to extract water from greater depths. Also, low cost generally encourages an excess water withdrawal, an inefficient usage pattern commonly exacerbated by ineffective application methods.

In India, it is estimated, in 2010, the total water withdrawal amounts to 761 km^3 of which 91% or 688 km^3, was for irrigation (FAO 2011). Also, in 2010, primary surface water withdrawal accounted for 52%, primary groundwater withdrawal for 33%, and reused agricultural drainage water for 15% (FAO 2011). Groundwater in India is a highly important resource for irrigation and household use, and its extensive use has resulted in widespread groundwater depletion (Shah et al. 2003; CGWB 2007). Irrigation progress in India can be traced back to ancient times. Early Indian scriptures talk of wells, canals, tanks and dams and their efficient functioning and preservation. Framji (1987) informs us irrigation to produce food grains was in existence for over 5000 years ago.

A study on groundwater depletion (Rodell et al. 2009) in the north-western Indian states of Haryana, Punjab, and Rajasthan, is illustrative of common regional groundwater depletion problems in India. Using the Gravity Recovery and Climate Experiment (GRACE) satellites for measuring changes in the terrestrial water storage during the study period from August 2002 to October 2008, 109 km3 of groundwater loss was estimated, or about 4 cm each year over the three-state area. A study by Shah (2011) finds, in respect of rural areas, electricity subsidies allowing farmers to pump groundwater cheaply becoming entrenched in the political landscape; and Low cost encourages an excess water withdrawal, an inefficient usage pattern commonly exacerbated by ineffective application methods (Kondepati 2011). This situation is perpetuated with further scarcity leading to unequal economic returns and, finally, takes on the most exploitative form where the "large landlords" also emerge as "water lords" through surplus accumulation, forcing the small and marginal landholders to become landless agricultural labourers (Sarkar 2011).

Wyrwoll P (2012), points out prominent drivers of over-extraction: inefficient usage, energy subsidies in agriculture, easy bank loan and population growth. A study by WRG (2009) suggests, in order to feed a growing and wealthier population, agricultural water demand in 2030 India would need double to 1200 billion m^3 if these inefficient practices continued. While from a resource management perspective, groundwater depletion could be said to be self-regulating; i.e., as groundwater is depleted, extraction becomes more expensive and groundwater withdrawals are reduced. There is little evidence to show that this self-regulation is happening in South Asia (Shah et al. 2006), and if it does, there are still severe consequences for households that lose the ability to irrigate their lands with groundwater sources.

A study carried out by Reddy (2005) in three villages in northern Andhra Pradesh that groundwater depletion costs are most damaging to farmers with smaller land holdings and that disparities in groundwater access and control generally develop – those having larger land holdings are able to adapt to groundwater depletion and still retain access to groundwater resources. A decline in water table gives rise to technological externalities in terms of rising costs of installing new wells, deepening of existing wells and pumping and other maintenance activities (Moench 1992 and Shah 1993). Shah et al. (2003) analyse the progression of groundwater use at four stages: (1) expansion of bore well installations; (2) groundwater-based agrarian boom; (3) onset of groundwater depletion concerns; and (4) collapse of groundwater-based systems. Observations from the Hoskote taluka study area (based on perceptions of farmers, number of failed/dry bore wells, and crop and occupational shifts) suggest that the study area is in between stages three and four. Obtaining more benefits from each drop of water consumed, especially from every drop irrigated agriculture consumes, can be the key to mitigating problems of scarcity.

Groundwater is a classic example of a 'public good' – a resource in respect of which, it is difficult to exclude the potential users and also more importantly, it is not in the self-interest of the individual to use the resource in a collectively beneficial manner, for if only, one or two users reduce the volume of water they withdraw, the overall impact will be minimal. In fact, all users compete with each other to extract as much water as they can while the resource still exists with the result that all find themselves more worse off than had they cooperated, reducing water consumption. This is an example of "Tragedy of Commons".

Since water resources are considered as resource-livelihood systems, it is beyond a water body, comprising a broader socio-cultural-ecological significance. They are the most dynamic life-support system having an enormous socio-economic-cultural-ecological importance to the communities depending on water resources for their livelihood sustenance. The communities have evolved their informal institutions (rules, norms) for managing the natural resource base (NR) like forests, fisheries and water resources with a view to sustaining the community life. There are multiple stakeholders associated with water management/use. But we selected only farmers having bore wells as our major stakeholder, as our main focus was to understand the ground water scenario.

Given this backdrop, the current study followed a two-level survey approach, namely, household level (especially bore well owners; running, failed and dried bore wells) and village level observations (to understand the village level dynamics). The study followed a selective sample for analytical purpose (Bore-well owners-present and past, water buyers, failed and dry bore well households). Besides the survey, at the village level, we collected information on the overall water resource dynamics. Under this, some specific village level information was collected such as details of bore wells and other irrigation sources, bore well depths, water levels, changes in the cropping pattern and perceptions of the villagers regarding the ground water situation. This apart, a few senior citizens also were interviewed for understanding issues on a temporal scale (how the situation was and is). Queries also focused on water availability around the villages (especially sources such as open wells, bore wells, rainwater storage tanks etc) and the ground water scenario in a historical context as well as changes that have taken place. We collected information from the same 102 sample bore well owning households (HHs), Among 102 farmers (HHs), 48 from Jadigenahalli, 15 from Govindapura, 10 from Vadigehalli, 16 from K B Hosahalli and 13 from Haraluru.

All the five study villages come under the semi-arid region of Karnataka with ground water as the only source for agriculture, domestic use and livestock activities. In the absence of other sources of water, there is a considerable pressure being exerted on ground water resource. As revealed by respondents, tanks have dried up since many years. For drinking water, about 80% of the households depend on piped water supply by the Government through *panchayat,* while about 54% of the households are dependent on the same source for livestock. The supply of water for domestic use is not sufficient with many disputes observed among the communities over water. For agriculture, 56% of the HHs have own bore-wells for irrigation. The average total land holding is over 3.5 acres, excepting Jadigenahalli where it is 4.17 acres. Irrigated landholdings have declined over the years due to the depletion of ground water and dried bore-wells with around 46% of the land holdings being irrigated.

Agriculture continues to remain the primary livelihood practice across the study villages. On an average, 60% of the households are engaged in agriculture as a major economic activity. However, dependency on agriculture has been declining over the years mainly due to ground water depletion. Other economic activities include livestock rearing, daily wage labour, business, company jobs etc. Among the secondary economic activities, livestock plays an important role. The fact that 19% of the households do not have any access to secondary occupations and that they are engaged in minor economic activities clearly indicates that agriculture as the major occupation is steadily declining due to water shortage.

Hosakote Taluk, one of the landscapes in the semi-arid tropical regions of Karnataka, depends only on groundwater for drinking, cultivation and industrial purposes. Groundwater levels have been declining with the over abstraction being a continuous phenomenon and currently it is one of the overexploited taluks with respect to natural replenishable quantity. Central Ground Water Board (CGWB) in its report on Bengaluru Rural District (2013) has mentioned that the status of development of groundwater in Hosakote Taluk is 223.49% i.e., overexploited. Studies related to Hosakote (Maruthesha Reddy et al. 2016) have indicated that the observation well in Upparahalli (near to Jadigenahalli GP) has been showing steady increase in the depth of ground water level below ground from 2005 to 2014. The depth to water level during pre and post monsoon (2014) has been mentioned as 34.54 and 31.66 m below the ground level as against 12.9 and 7.97 m during 1993. The exploratory drilling has (CGWB 2013) indicated the presence of water yielding fractures at 14.0 m, 108 m, 110 m and 123 m, 134 m, 136 m, 138 m depth, while the depth of exploratory bores drilled range from 33.0 to 287.09 m. The scientific exploration for potable water in the deeper zones has been going on with farmers testing their fortunes even at greater depths. While the scenario is overexploited and people have been reaching to deeper zones, distribution and depth relation at village level has not been attempted so far. Therefore, in order to understand the efforts of farmers to drawing water as part of coping with the vagaries of monsoons for cultivation, a census of groundwater abstraction structures was conducted as part of this study (Figs. 2.6, 2.7, 2.8, 2.9, 2.10, 2.11).

A detailed inventory under the Jadigenahalli GP (conducted during November 2017) area has revealed 480 groundwater withdrawal structures (Table 2.5). This includes both open-wells and bore-wells. Table 2.11 provides the distribution of these structures across different villages of the *Gram Panchayat*. Figure 2.12 shows the spatial distribution of groundwater withdrawal structures under Jadigenahalli

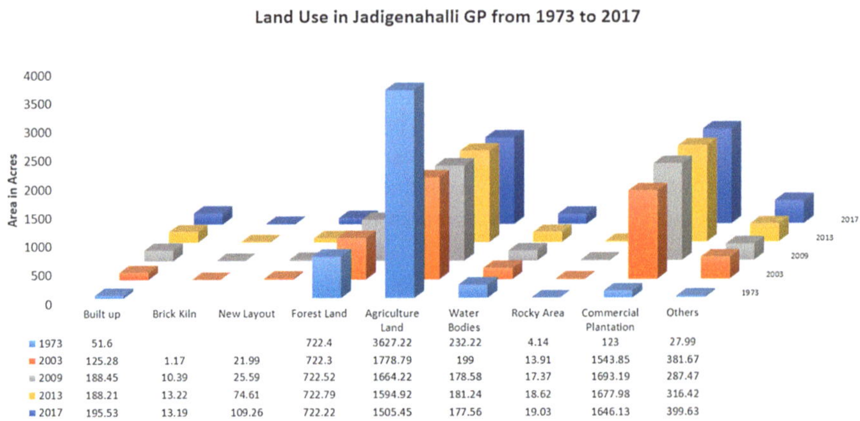

Fig. 2.6 Land use under Jadigenahalli GP from 1973 to 2017

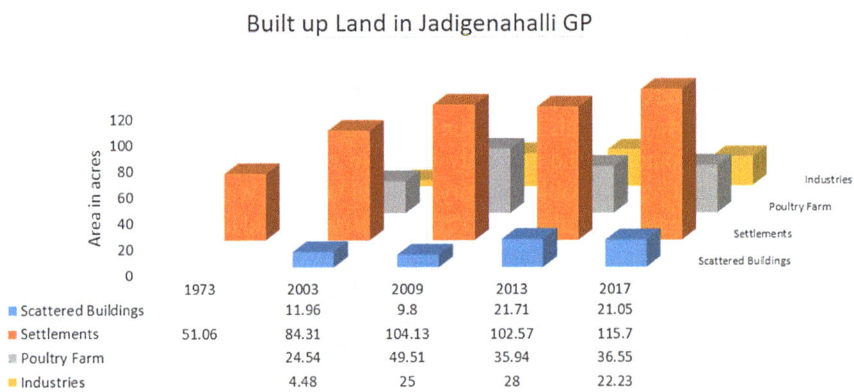

Fig. 2.7 Variations in the extent of built-up land under Jadigenahalli GP

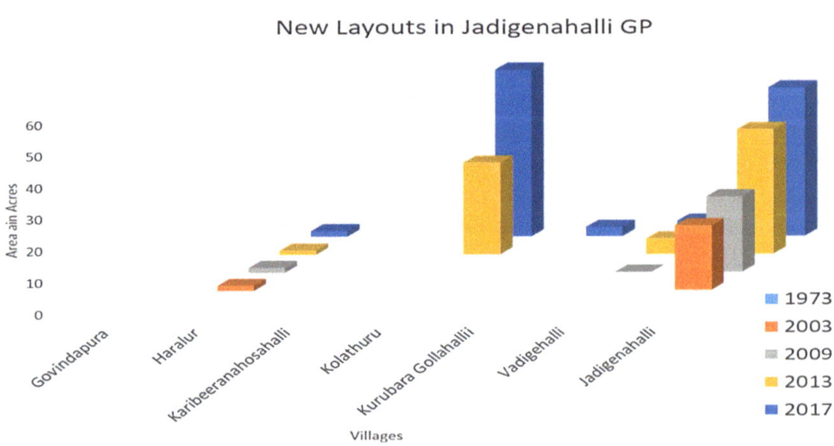

Fig. 2.8 Increasing trend of layouts under Jadigenahalli GP

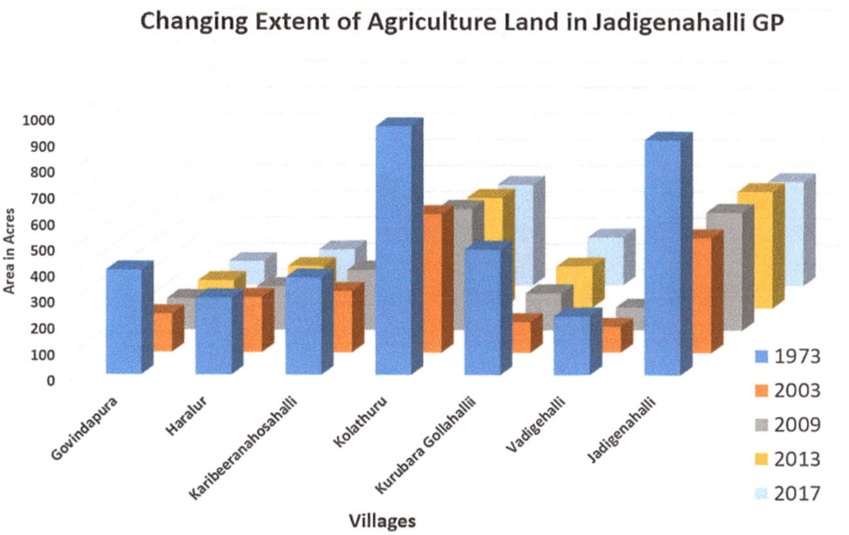

Fig. 2.9 Decreasing trend of agricultural lands under Jadigenahalli GP

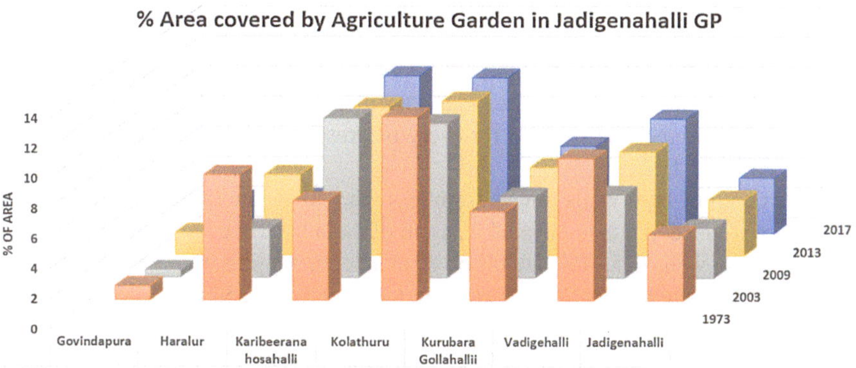

	Govindapura	Haralur	Karibeerana hosahalli	Kolathuru	Kurubara Gollahallii	Vadigehalli	Jadigenahalli
1973							
2003	0.9	8.3	6.6	12.2	5.9	9.4	4.3
2009	0.5	3.3	10.6	10.3	5.4	5.5	3.3
2013	1.5	5.4	9.8	10.2	5.9	6.9	3.7
2017	2.2	2.3	10.4	10.3	5.8	7.6	3.7

Fig. 2.10 Varying pattern of agricultural gardens under Jadigenahalli GP

Gram Panchayat. Few photographs taken at the site during the inventory are shown below (Plate 2.9, 2.10, 2.11, 2.12, 2.13, 2.14, 2.15 and 2.16).

It is observed that open wells have almost become obsolete with only bore wells being used. The survey has indicated that out of 67 open-wells 54 are dry (80.5%) and those which appear to be functioning run out of water soon after the monsoon season. The average depth of open-wells is around 35 feet below the ground level. A maximum wells are found at a depth ranging from 45 to 50 feet. Out of 413 bore wells, 144 (i.e., around 35%) have gone dry.

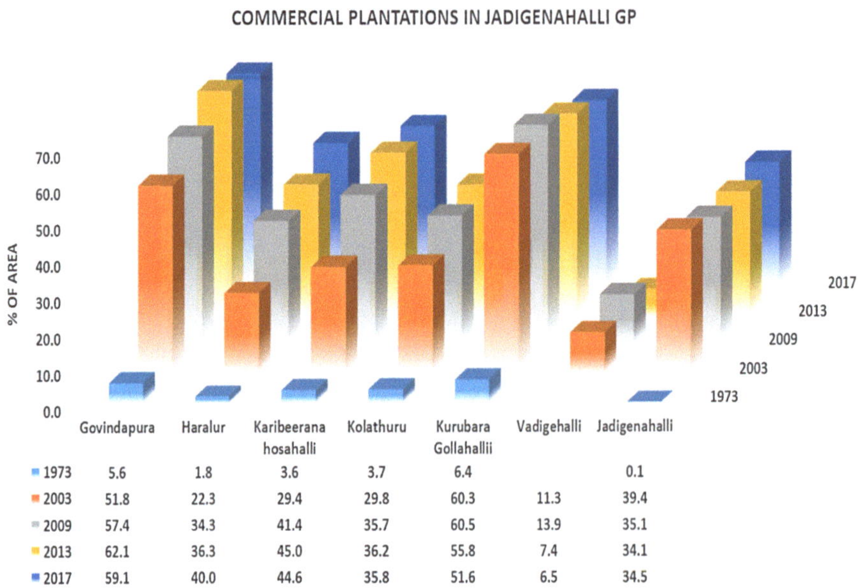

Fig. 2.11 Extent of eucalyptus (Commercial) plantations under Jadigenahalli GP

The study also indicates that the density of groundwater withdrawal varies from 6 acres to 35 acres per structure. The data indicate that Kolathur and Haraluru account for a higher density i.e., for every 6 acres, there is a withdrawal structure (Figs. 2.13 and 2.14).

The depth of bore wells ranges from less than 500 to more than 1500 feet. For 113 bore wells, depth is not available, as all of them have become defunct. A maximum of bore wells (25.91%) are observed at a depth ranging from 750 to 1000 feet (Annex 2.6), followed by 21.07% ranging from 1000 to 1250 and 12.59% ranging from 1250 to 1500 feet. The varying depths of bore wells indicate that people have kept on trying for water at deeper levels, currently beyond 1500 feet.

The data also indicates that more than 60% of the bore wells are deeper than 750 feet. Kolathur and Jadigenahlli account for a higher number of structures with a maximum withdrawal of groundwater. The percentage of defunct bore wells amounts to more than 30% with both Haralur and Kolathur sharing more number of defunct bore wells. Although bore wells are comparatively few in Karibeeranahosahalli, nearly 50% of them have become defunct.

Agriculture in this semi-arid area is highly dependent on groundwater irrigation with bore wells as the principal source of irrigation water. Considering that rain-fed agriculture apart, ground water extraction is the only source of irrigation, there is no surprise that the density of bore wells has increased manifold over the years. As a result, there is a severe depletion in groundwater, resulting in failed and dry bore wells. Discussions with bore well owners revealed that the maximum depth of bore wells has reached to 1300 fts (in 2014) but more than 1500 (in 2017 as indicated in

Table 2.5 Groundwater withdrawal structures under Jadigenahalli gram panchayat

Village name	Area in acre	Total structures	Density – Acre/per well	Bore well status			Open well status		
				Defunct	Functional	Total BW	Defunct	Functional	Total OW
Govindapura	465.94	32	15	9	20	29	2	1	3
Haraluru	314.18	56	6	15	23	38	13	5	18
Jadigenahalli	1025.83	123	8	31	84	115	8	0	8
Karibeeranahosahalli	430.7	55	8	25	28	53	1	1	2
Kolathur	1111.56	182	6	55	92	147	29	6	35
Kurubara Gollahalli	526.51	19	28	5	13	18	1	0	1
Vadigenahalli	461.12	13	35	4	9	13	0	0	0
Total		480		144	269	413	54	13	67

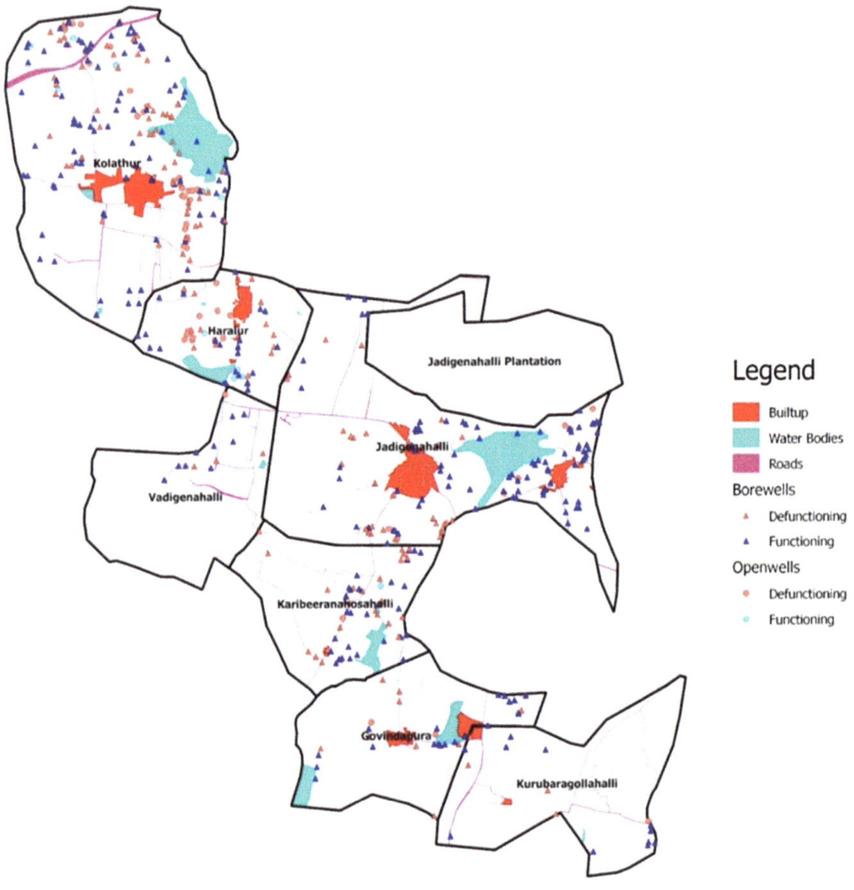

Fig. 2.12 Distribution of wells under Jadigenahalli gram panchayat

Fig. 2.15) indicating the gravity of the situation (Figs. 2.16 and 2.17). Figure 2.18 gives a schematic representation of increasing bore well depths over the decades as revealed by farmers.

2.3.3 Distribution of Bore Wells as Against Land Use

Obviously, bore wells under Jadignehalli GP are found across different land use categories. A maximum of them are found in agriculture lands (58.11%). 15.74% are seen in commercial plantations of which 53.85% of the bore wells are functional. Even people have drilled bore wells in tank beds. Besides, 56.67% of the drinking water wells in villages have also become defunct.

Plate 2.9 Old open well (currently not in use) near Kolathur

Plate 2.10 One of the open wells having good storage near Vadigehalli

The presence of bore wells in commercial plantations and “others” category of land use implies that these plantations were agricultural lands before being rendered to the current status of land use. The reasons for such a scenario could be: (1) farmers might have converted part of their agri land for raising eucalyptus plantations; (2) After bore wells became defunct, farmers might have opted for raising eucalyptus in their land; (3) Yields from bore wells might have dwindled/not sufficient for

Plate 2.11 Old water storage structure (Kalyani) near Jadigenahalli

Plate 2.12 Functional bore well providing irrigation to vegetables near Jadigenahalli

cultivating their lands, leaving portions of their land vacant which over the years have become scrublands. A detailed description of such implications is presented in the subsequent sections. Details of distribution, depth and the status of bore wells across different land use categories (Table 2.6) in respect of each village are shown in Annex 2.7 and 2.8 and the distribution and status of open wells shown Annex 2.9, bore well depth and land use relationship in Annex 2.10 (Table 2.7).

Based on the field survey and discussions held with farmers, 56% of the sample households have working bore wells with over 4 bore wells per household. While 27% of the bore wells have failed (water not found), over 55% have already dried up. Only 17.69% of the bore wells are found functional (Figs. 2.19 and 2.20). About 71–88% of the bore wells have either failed or dried up over the years. It is also

Plate 2.13 Functional bore well used for irrigation of maize near Kolathur

Plate 2.14 Unused bore well located in Scrubland near Haralur

observed that the success rate of bore wells is very insignificant. Table 2.8 and Fig. 2.21 provide details of bore wells, indicating the severity of the issue (Fig. 2.22).

A decline in water table gives rise to technological externalities in terms of rising costs of installing new wells, deepening of existing wells and pumping and other maintenance activities (Moench 1992 and Shah 1993). This observation supports

Plate 2.15 Defunct bore well Located in eucalyptus plantation (earlier Crop Land) near Haralur

Plate 2.16 Bore well drilled in tank bed near Kolathur

the scenario in this region as well. About 25 years ago, ground water depth was at about 180–250 ft. At present, each functional bore well costs about 3 lakh Indian rupees (equivalent to US$ 4700) There are multiple sources of bore well finance, such as banks and local money lenders (gold loan), including self finance. About 60% of the respondents have reported borrowing money (for drilling bore well) from banks, while 49% have reported investing personal money. Table 2.9 presents the investment pattern in respect of bore wells.

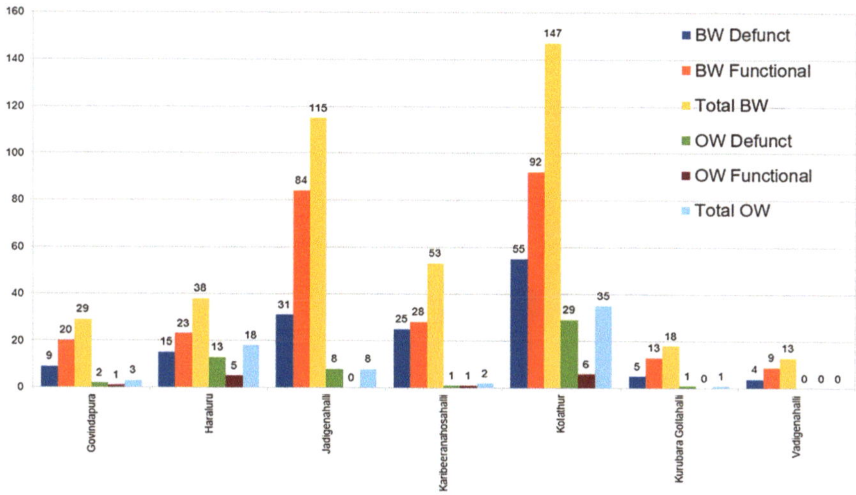

Fig. 2.13 Overall status of groundwater withdrawal structures under Jadigenahalli GP

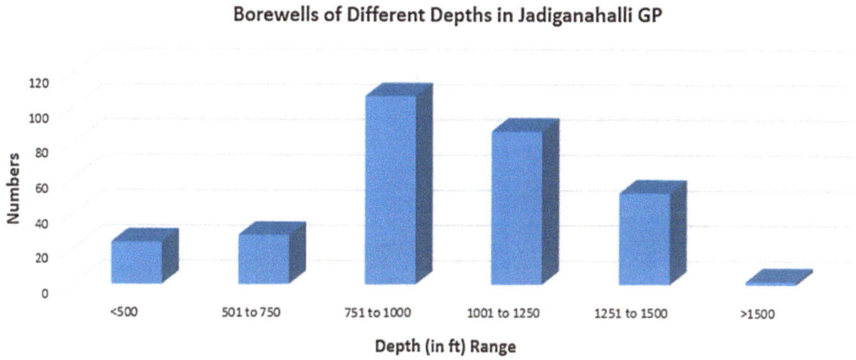

Fig. 2.14 Variations in the depth of bore wells under Jadigenahalli GP

An immediate effect of declining groundwater is reflected in changes in the cropping pattern with 62–90% of the farmers having changed their cropping pattern over the years.

2.3.4 *Farmers' Perceptions and Interventions*

Groundwater depletion is a major cause for concern in this water-stressed region, which is evident from over 80% of them perceiving the lack of rainfall as one of the main factors behind the drying up of bore wells and depletion of ground water besides pointing out four causal factors – (1) Scanty rainfall; (2) Over-extraction of

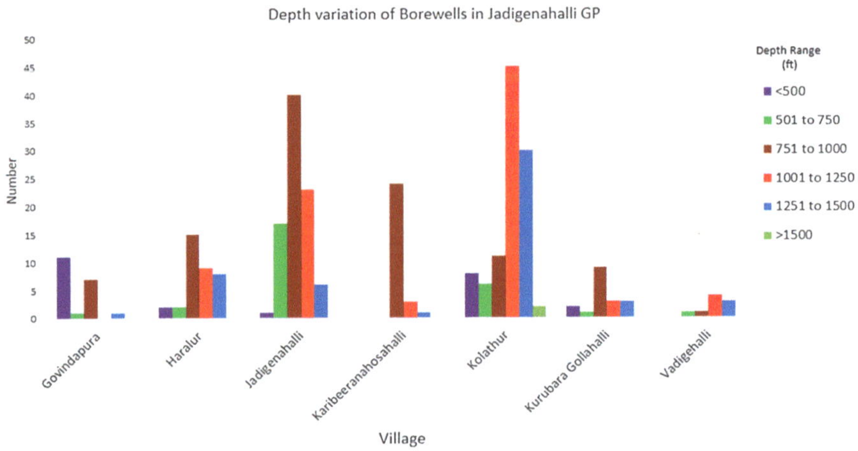

Fig. 2.15 Depths of bore wells in different villages under Jadigenahalli GP

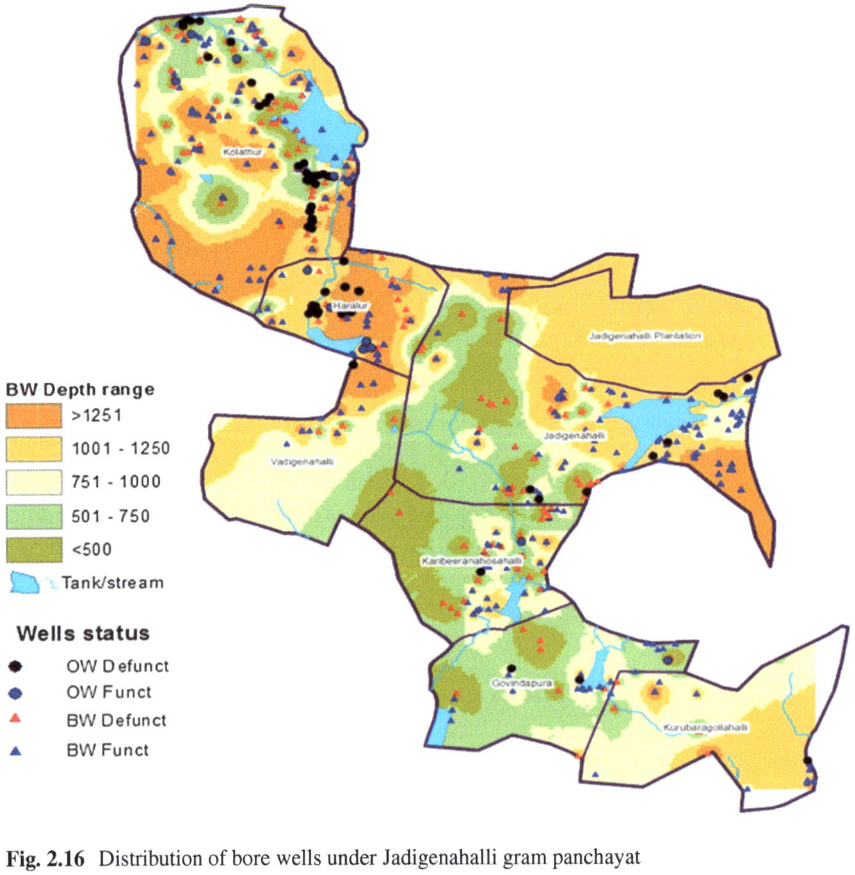

Fig. 2.16 Distribution of bore wells under Jadigenahalli gram panchayat

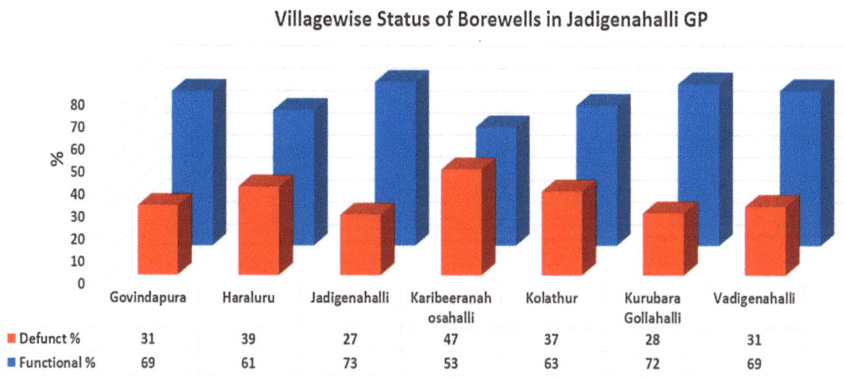

Fig. 2.17 Status of bore wells under Jadigenahalli GP

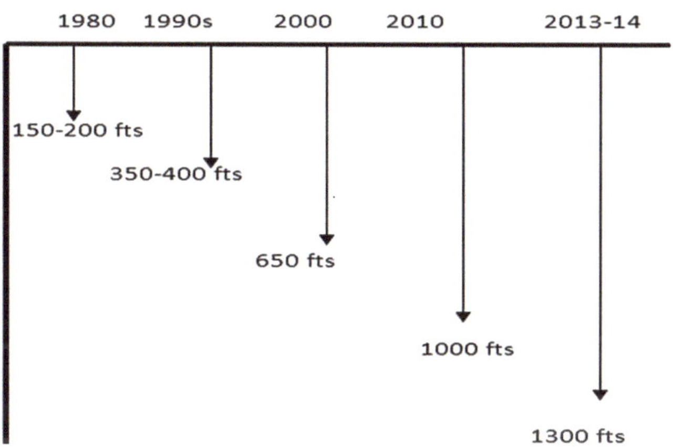

Fig. 2.18 Bore well depths indicated by farmers

ground water resources with an increase in the number of bore wells; (3) Degradation of Village tanks; (4) Deforestation. Respondents also have observed a positive relationship between the degradation of village tanks (no presence of water) and increasing number of failed and dried bore wells. 59% and 79% of respondents consider tank degradation and deforestation as major causes for declining groundwater tables respectively.

The increasing number of bore well is another reason for ground water depletion. In the earlier section, we have analysed that the increasing number of bore wells with an over-extraction of water is the major cause for ground water depletion, leading to the more number of failed and dried bore wells. About 70–80% of the respondents consider the increasing number of bore wells as the factor for the depletion of ground water. It has been observed in the field that Jadigenahalli and Haraluru have more number of bore wells. Figure 2.23 presents the views of respondents regarding the over-extraction of ground water and declining ground water levels.

Table 2.6 Location of wells in relation to land use

	Total bore wells		Status of bore wells						
Land use	Total	%	defunct	%	Functional	%	OWS Defunct	OWS Functional	Total
Agriculture land	240	58.11	73	30.42	167	69.58	31	10	41
Built-up	30	7.26	13	43.33	17	56.67		1	1
Commercial plantation	65	15.74	30	46.15	35	53.85	22	2	24
New layout	7	1.69	3	42.86	4	57.14			0
Others	54	13.08	20	37.04	34	62.96	1		1
Water bodies	17	4.12	5	29.41	12	70.59			0
Total	413		144	34.87	269	65.13	54	13	67

Table 2.7 Distribution of defunct open wells under Jadigenahalli GP

Village name	Total open wells	Agricultural land	Commercial plantation	Others
Govindapura	3	1	1	
Haralur	18	6	7	
Jadigenahalli	8	7	1	
Karibeeranahaosahalli	2	1		
Kolathur	35	15	13	1
Kurubara Gollahalli	1	1		
Total	67	31	22	1

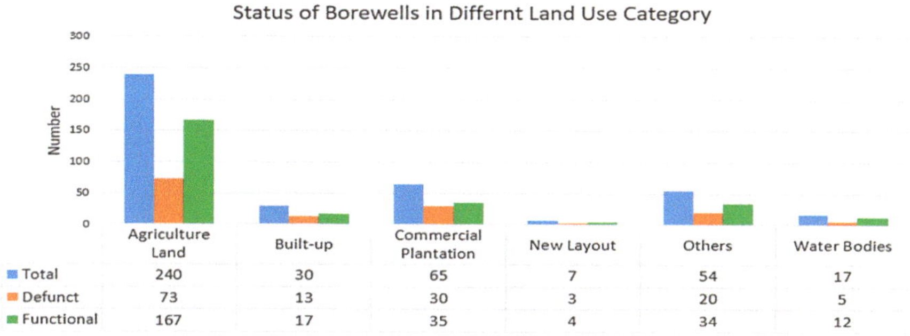

Fig. 2.19 Status of bore wells across different land use classes

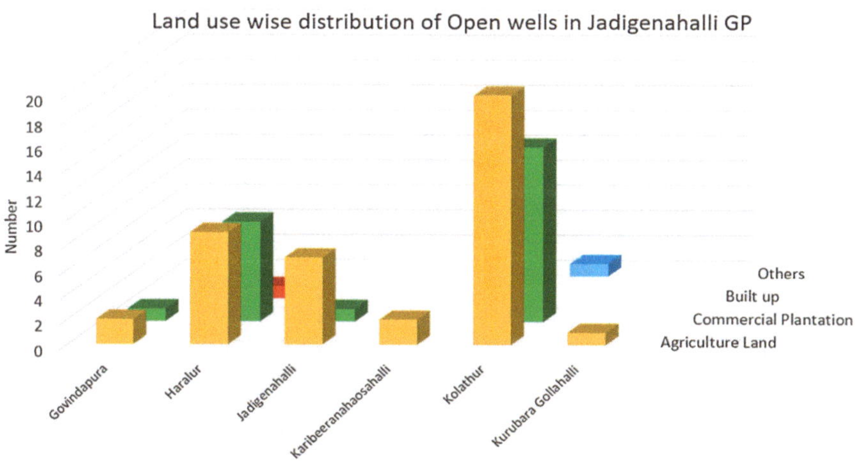

Fig. 2.20 Land use-wise distribution of open wells

Table 2.8 Status of bore wells as of 2014

Villages	Total drilled/ per HH	Failed	Dry	Running	% of failed bore wells	% of dry bore wells	% running	% of failed and dry bore wells
Jadigenahalli	5.29	1.46	2.90	0.94	27.56	54.72	17.72	82.28
Govindapura	1.87	0.60	0.73	0.53	32.14	39.29	28.57	71.43
Vadigehalli	5.00	1.30	3.10	0.60	26.00	62.00	12.00	88.00
KB Hosalli	2.63	0.88	1.13	0.63	33.33	42.86	23.81	76.19
Haralaru	5.15	1.00	3.46	0.69	19.40	67.16	13.43	86.57
All Villages	4.32	1.17	2.39	0.76	26.98	55.33	17.69	82.31

Source: Authors 2014

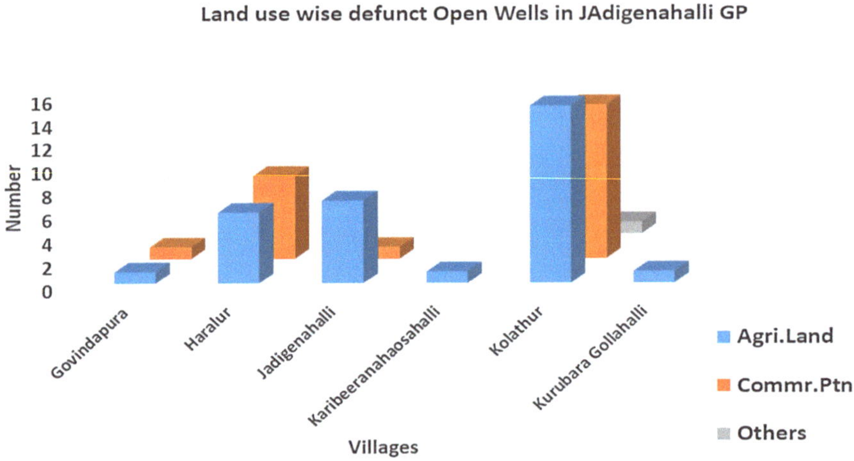

Fig. 2.21 Defunct open wells under Jadigenahalli GP

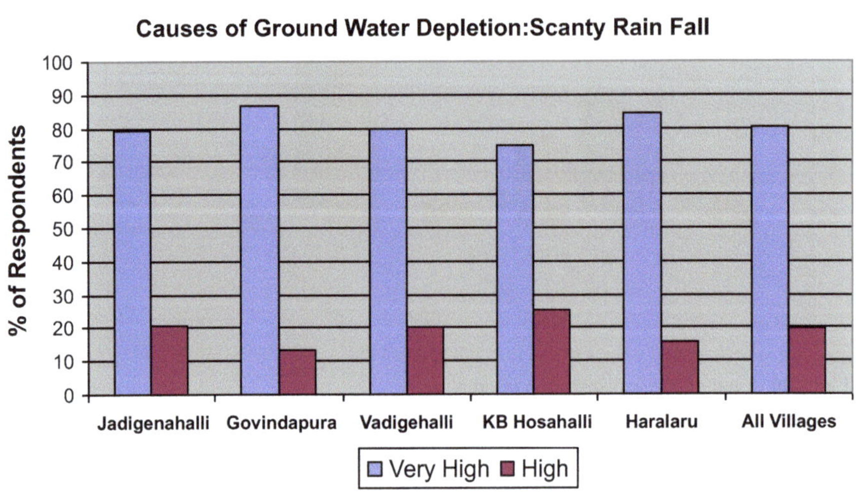

Source: Authors, 2014

Fig. 2.22 Ground water depletion: scanty rain fall

Table 2.9 Investment on functional bore-wells (Rs. Lakh)

Villages	Per BW	Per HH
Jadigenahalli	3.17	4.91
Govindapura	2.20	2.51
Vadigehalli	3.92	4.70
KB Hosahalli	2.57	2.86
Haralaru	2.78	3.57
All Villages	3.00	3.35

Source: Authors 2014

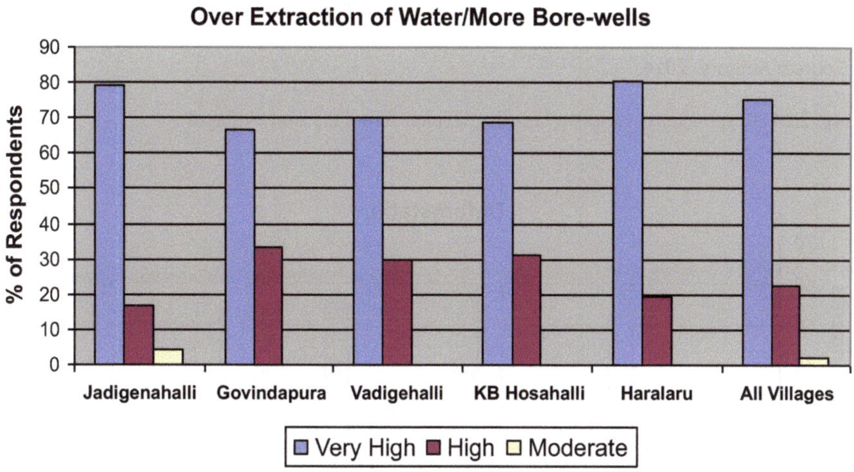

Source: Authors, 2014

Fig. 2.23 Increased number of borewells: over extraction of ground water

It is important to note that many respondents observe there is a positive relationship between the degradation village tanks (no water) and increasing number of failed and dried bore wells. In the study villages, 59% of the respondents consider (given a very high value) tank degradation as one of the major causes for declining ground water tables. In Jadigenahalli village, 69% of the respondents rate tank degradation as the main factor for the declining water table in and around the village. Figure 2.24 provides respondents views on village tank degradation and declining ground water table. Respondents have also observed that deforestation in the region is one of the major causes for groundwater depletion. Over 75% of the respondents have given a high to very high value to deforestation as one of the factors for the depletion of ground water (Fig. 2.25).

The consequences of groundwater depletion are grave (Fig. 2.26). The declining groundwater tables have posed significant challenges to the livelihoods of communities engaged in agriculture with over 40% of the households going through a

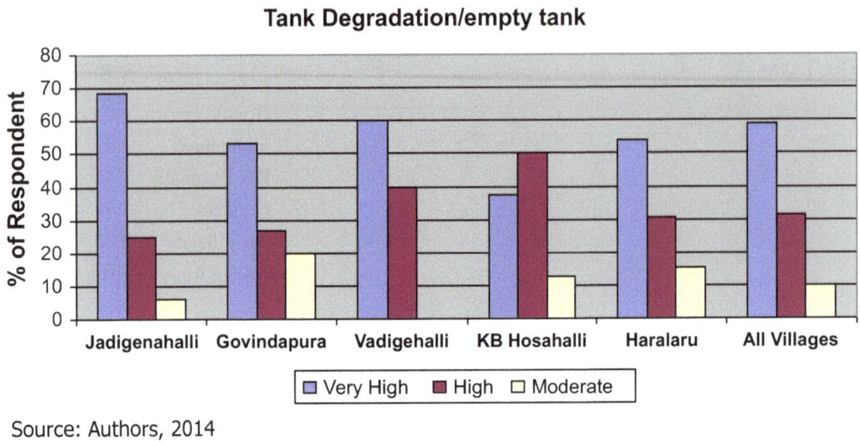

Fig. 2.24 Respondents' views on tank degradation across the study villages

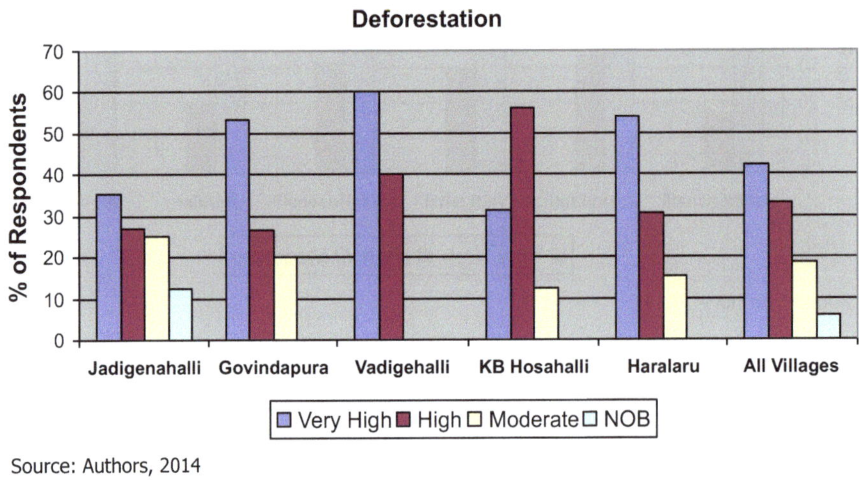

Fig. 2.25 Respondents' views on deforestation across the study villages

severe crisis and many households adopting alternative livelihood options due to a significant decline in the agricultural activities. Irrespective of owning agricultural lands, water scarcity has forced several of them (30%) to work as daily wage labourers. It is interesting to note that many HH have adopted alternative sources of livelihood (e.g.livestock rearing) in view of a significant decline in the agricultural activities due to groundwater depletion.

Focus group discussions and interviews with local farmers have indicated that they have strategized over the years to cope up with the changed scenario, from

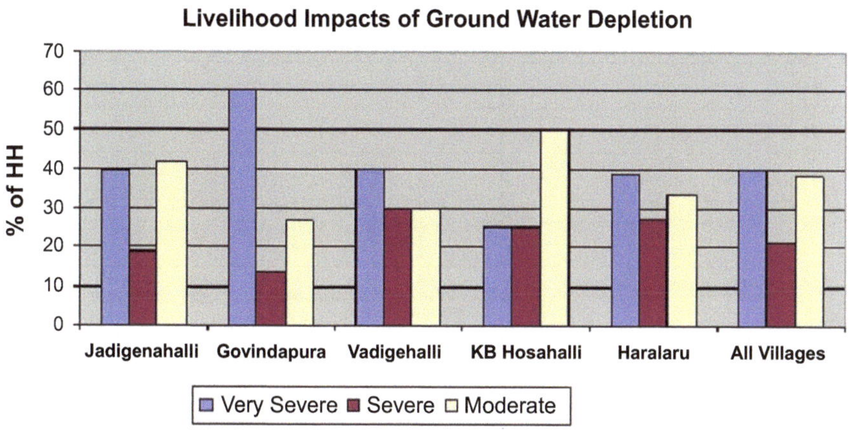

Source: Authors, 2014

Fig. 2.26 Livelihood impacts of groundwater depletion

being resource-rich to being resource-scarce. We have observed that unregulated groundwater exploitation, is a huge drain on household savings along with a forced cropping pattern shift from food grains to cash crops for easy cash flows and an inevitable out-migration in search of better jobs.

It is interesting to note that the sample respondents are well informed about the solution to the problem of groundwater depletion and implications for water scarcity. Their views towards with respect to addressing groundwater depletion and water scarcity issues are worth considering viz. – (i) Check dams in the region (68%); (ii) Rain water harvesting/recharging (57%); (iii) Afforestation and plantation in the region (not nilgiri plantation) (60%); (iv) Regulation of new bore wells; (v) Creating awareness for an efficient use of groundwater; (vi) Village tank restoration and management; (vii) River-tank water linkages; (viii) Diversion of Varthur lake water.

2.3.5 *Farmers' Coping Mechanisms*

Respondents' coping mechanisms suggestions for overcoming the problems of water scarcity and groundwater depletion are diverse. Over 86% of the total respondents have reduced their irrigated land area (already in two of the villages of Govindapura and Haraluru); more than 90% of the HHs have reduced their irrigated land; about 59% of the farmers have opted for additional bore wells; over 83% of farmers have changed their cropping pattern; 45% have adopted drip irrigation; 35% of the HHs have kept their lands barren partially or wholly due to dried bore wells and water scarcity; 45% of farmers have stopped commercial crop cultivation; 47%

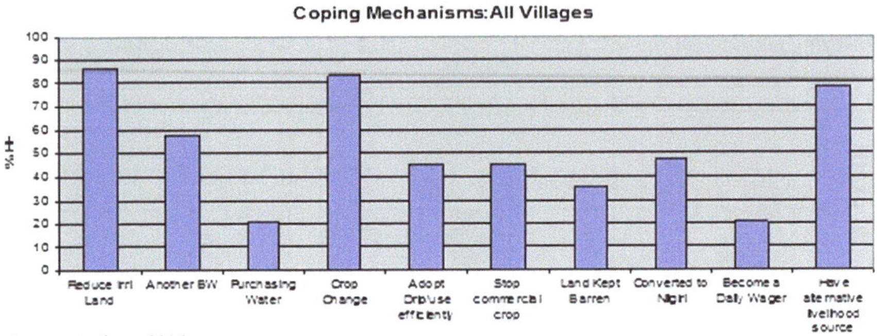

Source: Authors, 2014

Fig. 2.27 Coping mechanisms and adaptations

of the farmers have converted their agricultural lands into Eucalyptus plantations; 20% of the farmers have opted for daily wage labour; 78% of households have taken up alternative livelihood activities – viz., livestock rearing, company jobs, business etc. Fig 2.27 provides details of coping mechanisms adopted by communities in the context of declining groundwater and scarcity of water for irrigation.

We also have observed groundwater markets thriving that warrant a further examining of the property rights-related issues regarding groundwater use and management. Over 20% of the HHs purchase water from bore wells belonging to other farmers. Three types of water purchasing arrangements are observed across the villages – (i) monthly payment; (ii) purchasing water from tankers (for one tanker 350 rupees); and (iii) sharing of 25% of the value of agricultural products.

2.3.6 Adaptations to Water Scarcity: Technological Adaptation

2.3.6.1 Drip Irrigation

Lately, drip irrigation has received a considerable attention from policy makers, researchers and economists for its perceived ability to contribute significantly to groundwater resources development, agricultural productivity, economic growth and environmental sustainability. Over 45% of the households use drip technology for irrigation purpose, while Haraluru village accounts for the highest share (53.8%), followed by Jadigenahalli (52.1). It is observed that farmers engaged in grape cultivation use drip technology. Overall, drip irrigation accounts for about 66% of the total bore well based irrigation. Plate 2.17 and 2.18 shows cultivation of coriander and carrot using drip irrigation.

Source: Authors, 2014

Plate 2.17 Dried and abandoned open well and bore well

Source: Authors, 2014

Plate 2.18 Cultivation of coriander (left), carrot (right) using drip irrigation

2.3.6.2 Mulching

Mulching is practised as a coping mechanism for getting around water scarcity mainly for floriculture and growing of Rosemary (See Plate 2.19); mulch is a layer of material applied to the surface of a Soil area; it may be permanent (e.g. plastic sheeting) or temporary (e.g. bark chips). Its purpose is to conserve moisture and to improve fertility and health of the soil.

Source: Authors, 2014

Plate 2.19 Mulching done for growing rosemary, Jadigenhalli Village

2.3.7 Non-Technological Adaptation: Changing the Cropping Pattern

As discussed earlier, effecting changes in the cropping pattern seems to be a strategy followed by farmers as part of withstanding and overcoming water scarcity. Along with it, people have shifted from cultivating Paddy, Ragi and vegetables to raising Eucalyptus plantations, which are not labour-intensive and do not require external supply of water. Besides, there is a huge market for eucalyptus bark and leaves with brick and paper industrial units operating in the region (Mallur, Hoskote and Jadigenhalli Gram Panchayats), (See Plate 2.20).

2.4 Changing Ecosystem of Jadigenahalli – Push and Pull Factors

Development and degeneration are always associated with each other, but to what extent the degeneration can be sustained is determined by the response of the society and also the overall policy aspects. In the recent years, urbanisation has also

Source: Authors, 2014

Plate 2.20 Transporting of eucalyptus leafs for brick industrial units

been one of the main causes that have been exerting a substantial pressure on nearby rural ecosystems. Most of the rural landscapes in the semi- arid tropical regions and particularly the Peninsular India have been facing climate change challenges (mostly the vagaries of monsoons) as most of the river systems are dependent on rainfall. In addition to the climatic variations, urbanisation also has been one of the reasons for changing land use in this part of the country. In the regions surrounding Bengaluru, the situation is more pronounced, as there are no perennial rivers nearby. The livelihood depends on dry land agriculture. Bengaluru City, one of the fast growing metros, appears to have had effects of its growth on surrounding rural landscapes. Added to the urbanisation aspects, certain policy decisions like promoting eucalyptus might have also been a contributing factor to changing land use in rural areas. Almost all the surrounding rural areas of Bengaluru have been witnessing a rapid depletion of natural resources, particularly groundwater resources, while the land use has been getting transformed from agriculture to non-agriculture. Jadigenhalli is one of such samples of compounded policies, urbanisation compulsion and strategic actions of economic growth and response of the system.

Haroldo da Gama Torre (2011), while explaining the impact of urban sprawl worldwide, observes that the process is associated with the destruction and fragmentation of natural ecosystems, reduced diversity of species and loss of farmland. Priyadarshini Sen (2016) in relation to Hosakote and identifying the rural-urban fringe with characteristic land use associations opines that a conflict of two life styles and a direct impact of urban expansion on agricultural lands with clear indications of urban elements exist in peri-urban areas.

2.4.1 Agroforestry

There appear to have been certain policy decisions initiated by the government and practices followed by farmers that have influenced the ecological concerns of Hosakote region. The vagaries of monsoon and economic attractions might have prompted farmers' plant eucalyptus species in the agricultural lands. Due to the impact of different forestry development programmes, the farmers started growing some agro-forestry species, especially in the rural-urban areas. The Eucalyptus species were introduced in Karnataka State, India, during the year 1946, especially in the rainfed areas, as its uses are manifold and grows well in the low rainfall areas (Gangadharappa et al. 2003). The government also promoted fast growing Eucalyptus plantation species to cover the denuded areas, as part of afforestation programme. Eucalyptus was further introduced under World Bank aided Social Forestry project in the 1980s for firewood and small timber and was restricted to two districts, namely, Bengaluru (rural) and Kolar, replacing 70,000 ha under finger millet (ragi), a local staple food grain crop (Mukund Joshi and K. Palanisami 2011). Eucalyptus is a unique tree species as compared to other perennial trees, as regards its adaptability to water relations. It can efficiently adjust to surplus water situations with its water requirement rising to as high as 90 litres per plant per day. It can also successfully grow under water-scarce conditions with water requirements coming down to 40–50 l per plant per day. Unlike other perennial species, it is able to draw water from large area in the vicinity of its root system. Under stressful situations, its roots can spread deep even up to 6–9 m for extracting more water (Mukund and Palanisami 2011). Respondents indicated that eucalyptus was introduced in their village's way back in the 1980s by the Government with seedlings provided by the Forest Department. The farmers of Hosakote in Bagalore Rural district have continued to grow this easy-to-grow crop that requires less capital and attention. Jadigenahalli Gram Panchayat, as of 2017, has 1646.13 acres (34.38%) under Eucalyptus.

Earlier, a very few households were cultivating eucalyptus in villages, now majority have started cultivating because of low costs involved in plantation and maintenance (see annexure). Once in 3 or 6 years, it is harvested with cost incurred amounting to 4000–4500/ for acre. Not only almost one third of Jadigenahalli gram panchayat area is under eucalyptus, but also all the villages have more than 30% of the area covered by eucalyptus. Govindapura records highest coverage with 59.1% during 2017 while Vadigehalli has only 6.5% of the area under eucalyptus (See Table 2.10).

Spatial data on the geographical distribution of eucalyptus plantations for 1973–2003 was not available. However, based on the literature, it is clear that proliferation and cultivation of eucalyptus in agricultural lands must have begun during 1980–1990 and continued further. One of the studies reveals (Gangadharappa et al. 2003) that a majority of the farmers perceived the cultivation of eucalyptus species as being less expensive, including operational cost (78%). When profitability was analyzed, a maximum number of farmers (68%) felt that it had given higher profit and the year 1988 can be considered as the 'take off' year for the diffusion of eucalyptus.

Eucalyptus is a controversial tree species globally, due to its merits as well as notoriety. Its merits like fast growing nature, quick adaptation to a wide ranging

Table 2.10 Distribution of eucalyptus plantations under Jadigenahalli GP

Year	Govindapura	Haralur	Karibeerana hosahalli	Kolathuru	Kurubara Gollahalli	Vadigenahalli	Jadigenahalli	Total of Jadigenahalli GP
1973	25.99	5.71	15.48	41.55	33.54		0.71	122.98
2003	241.54	69.97	126.68	331.26	317.44	52.28	404.68	1543.85
2009	267.64	107.77	178.32	396.87	318.54	64.21	359.84	1693.19
2013	289.26	114.19	193.96	402.8	293.54	34.32	349.9	1677.98
2017	275.35	125.64	192.14	397.92	271.72	29.84	353.52	1646.13

ecological situations, several industrial applications and as a means of livelihood for the unprivileged, have elevated it to one of the most desirable tree species introduced in afforestation, farm forestry and social forestry programmes. Nevertheless, Eucalyptus is also known to cause a number of environmental hazards like depletion of groundwater, dominance over other species through allelopathic effects, loss of soil fertility and negative impacts on local food security issues. (Mukund and Palanisami 2011). The presence of groundwater withdrawal structures in the eucalyptus plantations (on private lands) corroborates the preference of farmers for eucalyptus as a means to reduce the cost of cultivation and to get a better price for the produce. The current study has been able to map both bore wells and open wells across the commercial plantations and "others" (cultivable waste – scrubland) category of land use from all the villages under Jadigenahalli GP. The scenario can be interpreted in the following ways:

1. Bore wells installed on a portion of the land becoming increasingly dry with no sufficient water available, and hence farmers have started cultivating eucalyptus in the same land (See Table 2.11).
2. Farmers must have reduced the extent of bore well irrigated land so as to reduce the cost of labour and to support other better remunerating crops. As a result, a substantial proportion of land with functioning bore wells is seen with eucalyptus crop in a few villages under Jadigenahalli GP.

Table 2.11 Village-wise and land use-wise distribution of wells under Jadigenahalli GP

			Bore wells		Open well	
Sl. No.	Village name	Land use class	Defunct	Functional	Defunct	Functional
1	Govindapura	Commercial plantation	1	4	1	
		Others	3	3		
2	Haraluru	Commercial plantation	6	5	7	1
		Others		3		
3	Jadigenahalli	Commercial plantation	5	8	1	
		Others	3	13		
4	Karibeerana hosahalli	Commercial plantation	3	6		
		Others	4	3		
5	Kolathur	Commercial plantation	14	11	13	1
		Others	7	9	1	
6	Kurubara Gollahalli	Commercial plantation	1	1		
		Others	1	1		
7	Vadigehalli	Others	2	2		
Total			50	69	23	2

3. Yields from bore well becoming increasingly dwindled/not sufficient for the entire lands – to reduce the cost of cultivation, farmers must have kept portions of their lands (with bore well/open well) unused which over the years have become scrublands.

It is also observed that more than 50% of the bore wells are still functional, providing water to adjacent crop/garden land.

2.4.2 Status of Environment in Jadigenahalli

The scenario of changing land use and overexploitation of groundwater in Hosakote Region in general and Jadigenahlli in particular can be analysed using the **Pressure-State-Response** model of OECD (2003). Following sections describe briefly the pressure, status of environment and response of society (citizen and institutional and governmental).

2.4.3 Natural Resources Stock – Land Use and Groundwater

Overall land use pattern under Jadigenahllli GP has transformed from agriculture-dominated to non-agriculture with agroforestry accounting for the lead portion. Agricultural land use has diminished substantially from 3627 acres to 1505 acres over a period of almost 40 years. At the same time, the eucalyptus (agroforestry) area has increased from 123 acres to 1646 acres. While one cannot determine yard-sticks for a standard/optimal land use in respect an agrarian landscape, particularly the one like Jadigenahalli GP, impacts of a modified land use practice is clear. Almost one third of the geographical area is under eucalyptus plantations raised intentionally to improve the economic status. The plantations might have possibly improved the economic conditions of farmers temporarily, but the same have, in the process, affected adversely the local landscapes. Eucalyptus has been at the centre of a number of controversies like loss of soil fertility, adverse effects on local food security, replacement of conventional forests, besides being a prime cause behind various hydro-ecological imbalances of local eco-systems. Experiments have shown that organic matter and exchangeable potassium have depleted in the soil systems under eucalyptus plantations than in wood lands (George 1986).

Studies have revealed that eucalyptus is prone to a larger suction in the upper zone, seriously reduce the natural movement of water before ultimately reducing the supply to deeper aquifers (Mukund Joshi and Palaniswamy 2011) and water use based on over three dry years of measurement, amounting to more than the rainfall (Calder et al. 1997). The study by Mukund Joshi and K Palanisami (2011) on the ground water conditions in Kolar area reveals that 20 years of continuous cultivation of eucalyptus in private and public lands had deepened the freshly-dug bore wells up to 260m from a mean depth of 177m. A survey of distribution and status of bore wells in Jadigenahalli (during November 2017 by authors), has clearly brought out the effects of eucalyptus on ground water levels. Majority of the depths of bore wells have increased more than 750 feet.

Groundwater drafting in Hosakote has been estimated at 223% by CGWB. As there are no perennial streams/rivers in the region, villages under Jadigenahalli GP have a great concern to deal with regarding water not only for agriculture, but also for drinking water, if this status continues.

2.5 Societal Responses

Urbanisation is inevitable in the context of a mounting pressure on land, reduced agricultural lands/incomes, increased demand for drinking water for the burgeoning population and also for other uses. Although urbanisation symbolises development, growth and change, an uncontrolled urbanisation process is creating substandard living environments, acute shortage of public services and environmental degradation around cities called peri-urban spaces. Jadigenahalli, one similar peri-urban landscape around Bengaluru city, has come to be impacted substantially by urbanisation and related developmental activities. On one side, the scenario has been moving away from containment options, despite efforts being made continuously towards building resources and controlling a further land degradation. On the other hand, there are coping mechanisms adopted by both landholders and landless across the villages. However, there is also the issue of neglect of common property resources. The responses at the societal level can be briefly mentioned as follows:

- Formation of BMRDA and Local Planning Authorities (LPAs) for regulation, coordination of urban development and land conversion to non-agriculture purposes.
- Bengaluru and its hinterlands have evolved and transformed over the years subject to the influences of various factors affecting the City–Region dynamics. The new millennium is being increasingly referred to as one Urban–Rural Continuum in India. Realising that regional planning is a tool to address regional disparities and that interplay between Bengaluru (Metropolitan) and its human settlements (surrounding urban and rural local bodies) would largely govern the State Level spatial and economic development, BMRDA has prepared a Revised Structure Plan for Bengaluru Metropolitan Region (Revised Structure Plan –2031).
- Hosakote Planning Authority (HPA) has prepared a Master Plan for its local planning area, which includes entire the Hosakote Taluk. Besides focussing on urban issues, the plan aims to address the issues like Zoning of Land use in rural regions, rural infrastructure, environment, regulation of groundwater exploitation and rainwater harvesting (Anonymous 2013).
- An official ban on cultivation and propagation of eucalyptus species by anybody across Karnataka by amending the Karnataka Tree Preservation Act 1976 (Anonymous 2017). This action by the government may go a long way in facilitating the percolation of rain water into groundwater aquifers that have been almost not receiving any rainfall percolation from soil zone due to the pres-

ence of eucalyptus plantations. The ill effects of growing eucalyptus appear to have been understood by the farmers as well, which is reflected in reducing the area under these plantations across different villages under Jadigenahalli GP.

- Farmers adapting to drip and sprinkler irrigation for crops in their lands. The survey during November 2017 indicates that many farmers have stopped flood irrigation and have instead resorted to an advanced method of irrigation, thereby reducing slightly the land under groundwater irrigation.
- Initiatives towards building water resources in Hosakote Region – Indirect reuse of waste water from Bengaluru.
- Most of the tanks in Hosakote taluk have gone dry with water tables dipping further. The normal monsoons are not able to bring back the increased demand. The Government has initiated plans to pumping of treated wastewater from the lakes around Bengaluru city to other regions like Kolar, Hosakote and Chikkaballapur. Bengaluru Mirror Bureau (2015) in an article based on a discussion with Bengaluru Water Supply & Sewerage Board (BWSSB) authorities and Dr. A N Yellappa Reddy (environmentalist) mentioned the following: At present, Bengaluru generates 1500 MLD of sewage every day, out of which 750 MLD is treated by the Bengaluru Water Supply and Sewage Board (BWSSB). This treated water is not fit for drinking. If all of Bengaluru's sewage is treated, 500 MLD could be used within the city and remaining can be sent to lakes in neighbouring districts. If the treated water is made to flow for 5 km on ground rich in vegetation and bio-diversity, all toxins in the water are removed naturally. Water could then be supplied via tunnels, and lifted from an ideal location to water bodies in neighbouring districts. Treated water supplied to tanks can be subjected to further purification so that people can drink it.
- Hosakote lake was probably the first lake in the vicinity of Bengaluru to have been filled with treated water. Water from Yelemallappa shettyhalli Kere near Avalahalli is being pumped to a distance of 14 km through a pipeline to Hosakote Lake for the last 5 years. Residents of Hosakote town say that treated water has done wonders to the water table. The water table, which had depleted to 1200 feet 5 years ago, has now come up to 650 feet (within a 5 km radius of the lake). Surprisingly, 20 bore wells have been dug up in the lake to supply drinking water to Hosakote town (Bengaluru Mirror Bureau – 2015).
- As the forest resource is under the control of the Forest Department, the dependency rate of people has got reduced extensively, though there is a theft of resources happening, but rarely reported.
- A major part of the grazing lands has been given to economically backward communities. They grow finger millet and Eucalyptus. The rest of the managed grazing land remains an open access (un-managed) CPR and hence, prone to degradation. Poultry farms and brick making industrial units have encroached upon grazing lands across the study villages.
- For the last 10 years and more, no tank is used for irrigation, as there is no water stored. In Govindpura, the tank bed has been used for SC community settlement. Many people continue to graze their goats/sheep in the tank bed. Silt removed from tank beds is auctioned to brick making industrial units operating around the villages.

- There is a dilution of village-level informal institutional arrangements pertaining to livelihood-natural resource based CPRs. But the causes of CPR degradation are external to the community-CPR inter-linkages. As there is no water present in the tank, the dependency level is less. Silt removal by brick making industrial units is also external to the village community. This could explain as to how external and physical shocks (like Water scarcity) have led to the breakdown of village institutions pertaining to the CPRS. Now all HHs have cross-breed livestock with a relatively low dependence on CPRs (high profit motive and degradation of CPR).

2.6 Summary

A major part of the grazing lands has been given to economically backward communities. They grow ragi and Eucalyptus. The rest of the managed grazing land remains an open access (un-managed) CPR and hence, degradation. Poultry farms and brick making industrial area have encroached upon grazing lands across the study villages. There has been a decline in livestock rearing, because earlier, people used to take their livestock to Gomalas, or waste lands with grass and shrubs for grazing but now, they have been converted into agriculture lands. Interestingly, the population of ox and buffaloes is less compared to cows, sheep and goats. Especially with respect to cows, people have shifted from local (hallikar) to hybrid cows (Holo frestine, jersy) because of their high milk yield (Manasi et al. 2016). As the forest resource is under the control of the forest department, the dependency rate of people has got reduced extensively though there is a theft of resources happening now and then (rarely reported).

For the last 10 years and more, no tank is used for irrigation as there is no water. In Govindpura, the tank bed has been used for SC community settlement. Many people continue to use to graze their goats/sheep in the tank bed. Silt removed from tank beds is auctioned to brick making industrial area operating around the villages. There is a dilution of village level informal institutional arrangements pertaining to livelihood-natural resource based CPRs. But the causes of CPR degradation are external to the community-CPR inter-linkages. As there is no water present in the tanks, dependency level is less. Silt removal by brick making industrial units is also external to the village community. This could explain as to how external and physical shocks (like Water scarcity) have led to the break-down of village institutions pertaining to the CPR. Now all HHs have cross breed livestock with less dependence on CPRs (high profit motive and degradation of CPR).

As per people's perceptions eucalyptus is the main reason for a rise in temperatures in the recent years; also no crops will grow once it is uprooted including the surrounding lands because it absorbs soil moisture from the surrounding areas. Since Jadigenahalli is just about 40 Km away from the metropolitan city, Bangalore, with a good communication and transportation network with NH4 (National Highway) connection and also as it possesses a forest area of 750 acres with natural resource access, the region less vulnerable to climate change effects (as compared to the remote region/villages). Now with only 25–26% of the villagers dependent on agriculture as their major occupation and the remaining 60–70% of the people on garments, factories, non-agriculture labour, livestock rearing as major alternative sources of their livelihood, they are less vulnerable to climate change. In view of this change, 60% of the people migrate daily to other cities in search of employment and most of the youngsters migrate daily in pursuit of education and thus changes in livelihood and migration are not due to climate change in this region.

Assessment of climatological indicators shows changes in rainfall pattern as well as temperature in the area. Since the data is constrained by lack of historical samples, climate change can be conclusively proclaimed only after such historical samples are examined or after monitoring for few more years. However, the studied data do not show distinct signs of climate change. Indirect indicators such as change in cropping pattern and agricultural productivity have been assessed. Change in cropping pattern from crops to eucalyptus plantation in study area can be attributed to lower ground water table due to excessive bore well drilling and economic profitability of eucalyptus plantation. However quality and productivity of grape and mulberry cultivation could have been affected by weather change. Concrete inference in this area could be drawn only after detail monitoring. Changes in demographic and economic indicators have also been attributed to change in cropping pattern and shift in employment pattern (Manasi et al. 2016).

The largest known economic impact of climate change is upon agriculture because of the size and sensitivity of the sector, particularly in the developing world and to a lesser extent in parts of the developed world. Even though there is a decline in agriculture activity in all the villages, but secondary data shows that percentage of area under agriculture in all villages, expect Vadigehalli and KB Hosahalli have increased. Technical indicators like rainfall and temperature indicate that there is no climate change in the region and as due to basic necessities of village are good and accessible along with superior natural vegetation has made the region less prone to climate change effects (Manasi et al. 2016).

A well-planned groundwater resource management strategy is important for an economical, efficient and sensible use of ground water. In view of the ever-growing population and increasing demand for groundwater for various developmental activities, it is suggested to adopt methods to artificially recharge ground water aquifers in the water level depleting areas, as part of increasing the ground water availability. Apart from the above, farmers should be encouraged to grow crops that require less water for their production. With respect to ground water resources management, some interesting suggestions made by farmers include the construction of check dams in the region; Rain water harvesting/recharging; Afforestation and plantation in the region (not nilgiri plantations); Regulation of new bore wells; Creating awareness for an efficient ground water use; Village tank restoration and management; River-tank water linkages; Diversion of Varthur lake water to the region.

Annexures

Annex 2.1 Current land use pattern (2017) under Jadigenahalli GP

Land use classes	Govindapura	Haralur	Karibeerana hosahalli	Kolathuru	Kurubara Gollahallii	Vadigehalli	Jadigenahalli	J. Plantation	Total of Jadigenahalli GP
Built-up	22.60	19.78	3.78	69.08	6.29	4.90	69.10		195.54
Brick kiln	1.33	0.95	2.24	6.16	2.51				13.19
New layout		1.77		52.79	2.97	4.88	46.84		109.26
Forest (land) plantation				38.40		231.70		452.16	722.26
Agriculture land	91.82	138.54	187.64	386.88	185.05	115.59	399.89		1505.40
Water bodies	23.48	13.95	12.61	69.25	0.51	0.76	56.99		177.56
Rocky area	0.11	1.07		7.09	10.76				19.03
Commercial plantation	275.35	125.64	192.14	397.92	271.72	29.84	353.52		1646.13
Others	51.25	12.49	32.28	83.98	46.70	73.45	99.48		399.63
Total area	465.94	314.18	430.70	1111.56	526.51	461.12	1025.83	452.16	4788.00

Annex 2.2 Land use pattern (2013) under Jadigenahalli GP

Land use classes	Govindapura	Haralur	Karibeerana hosahalli	Kolathuru	Kurubara Gollahallii	Vadigenahalli	Jadigenahalli	J. Plantation	Total of Jadigenahalli GP
Built-up	16.91	19.35	2.65	68.47	6.02	10.62	64.19		188.21
Brick kiln	2.25		1.23	6.16	3.58				13.22
New layout		1.33		29.02		4.89	39.37		74.61
Forest (land) plantation				38.40		234.54		452.16	725.10
Agriculture land	102.92	158.18	196.64	422.22	157.95	107.70	447.00		1592.61
Water bodies	21.07	14.07	19.00	67.72	0.51	0.82	58.05		181.24
Rocky area	0.05	1.16		6.72	10.69				18.62
Commercial plantation	289.26	114.19	193.96	402.80	293.54	34.32	349.90		1677.98
Others	33.48	5.95	17.21	70.01	54.21	68.23	67.32		316.42
Total area	465.94	314.23	430.70	1111.53	526.51	461.12	1025.83	452.16	4788.01

Annex 2.3 Land use pattern (2009) under Jadigenahalli GP

Land use classes	Govindapura	Haralur	Karibeerana hosahalli	Kolathuru	Kurubara Gollahallii	Vadigenahalli	Jadigenahalli	J. Plantation	Total of Jadigenahalli GP
Built-up	14.26	17.34	2.20	71.18	4.60	9.20	69.67		188.45
Brick kiln	2.46		2.92		5.01				10.39
New layout		1.58				0.12	23.89		25.59
Forest (land) plantation				38.40		233.66		452.16	724.22
Agriculture land	119.68	166.23	229.55	464.81	141.99	86.99	453.27		1662.53
Water bodies	19.76	13.12	7.13	77.70		0.66	60.22		178.58
Rocky area	0.47	1.61		5.26	10.00	0.03			17.37
Commercial plantation	267.64	107.77	178.32	396.87	318.54	64.21	359.84		1693.19
Others	41.68	6.53	10.58	57.33	46.35	66.25	58.74		287.47
Total area	465.94	314.18	430.70	1111.56	526.49	461.11	1025.63	452.16	4787.78

Annex 2.4 Land use pattern (2003) under Jadigenahalli GP

Land use classes	Govindpura	Haralur	Karibeerana hosahalli	Kolathuru	Kurubar gollahallii	Vadigenahalli	Jadigenahalli	J. Plantation	Total of Jadigenahalli GP
Built up	14.59	8.71	2.91	43.51	2.77	7.34	45.45		125.28
Brick kiln			1.17						1.17
New layout		1.63					20.35		21.99
Forest (land) plantation				38.40		231.74		452.69	722.83
Agriculture land	145.93	210.70	234.92	531.35	117.34	98.18	440.37		1778.79
Water bodies	24.85	12.99	25.26	71.23	0.51	1.60	62.56		199.00
Rocky area		1.25		2.29	10.38				13.91
Commercial plantation	241.54	69.97	126.68	331.26	317.44	52.28	404.68		1543.85
Others	39.04	8.92	39.75	93.52	78.02	69.99	52.43		381.67
Total area	465.94	314.18	430.69	1111.56	526.47	461.13	1025.83	452.16	4787.96

Annex 2.5 Land use pattern (1973) under Jadigenahalli GP

Land use classes	Govindpura	Haralur	Karibeerana hosahalli	Kolathuru	Kurubar gollahallii	Vadigenahalli	Jadigenahalli	J. Plantation	Total of Jadigenahalli GP
Built-up	6.65	2.49	4.55	8.42	1.43		27.52		51.06
Brick kiln									0.00
New layout									0.00
Forest (land) plantation				38.40		232.12		452.16	722.68
Agriculture land	400.28	294.51	372.18	953.37	481.20	224.66	900.77		3626.97
Water bodies	32.13	11.44	38.48	56.21	0.21	3.11	90.63		232.22
Rocky area					4.14				4.14
Commercial plantation	25.99	5.71	15.48	41.55	33.54		0.71		122.98
Others	0.91	0.02		13.61	5.96	1.25	6.21		27.96
Total area	465.95	314.18	430.70	1111.56	526.48	461.14	1025.83	452.16	4788.01

Annex 2.6 Depth (variation) of bore wells under Jadigenahalli GP

Village	Depth of bore wells (in feet)							Total	%
	Depth not available	< 500	501–750	751–1000	1001–1250	1251–1500	> 1500		
Govindapura	9	11	1	7	0	1	0	29	7.02
Haralur	2	2	2	15	9	8	0	38	9.20
Jadigenahalli	28	1	17	40	23	6	0	115	27.85
Karibeeranahosahalli	25	0	0	24	3	1	0	53	12.83
Kolathur	45	8	6	11	45	30	2	147	35.59
Kurubara Gollahalli		2	1	9	3	3	0	18	4.36
Vadigehalli	4	0	1	1	4	3	0	13	3.15
Total	113	24	28	107	87	52	2	413	100.00
% of bore wells	27.36	5.81	6.78	25.91	21.07	12.59	0.48	100	

Annex 2.7 Village-wise and land use-GP wise distribution of wells under Jadigenahalli GP

Sl. no.	Village name	Land use class	Bore well status		Open well status	
			Defunct	Functional	Defunct	Functional
1	Govindapura	Agriculture land	3	11	1	1
		Built-up	2	1		
		Commercial plantation	1	4	1	
		Others	3	3		
		Water bodies		1		
		Sub total	9	20	2	1
2	Haraluru	Agriculture land	9	13	6	3
		Built-up		1		1
		Commercial plantation	6	5	7	1
		Others		3		
		Water bodies		1		
		Sub total	15	23	13	5
3	Jadigenahalli	Agriculture land	20	54	7	
		Built-up	1	3		
		Commercial plantation	5	8	1	
		New layout	2	2		
		Others	3	13		
		Water bodies		4		
		Sub total	31	84	8	0
4	Karibeeranahosahalli	Agriculture land	16	18	1	1
		Built-up	1	0		
		Commercial plantation	3	6		
		Others	4	3		
		Water bodies	1	1		
		Sub Total	25	28	1	1
5	Kolathur	Agriculture land	22	56	15	5
		Built-up	7	9		
		Commercial plantation	14	11	13	1
		New layout	1	2		
		Others	7	9	1	
		Water bodies	4	5		
		Sub Total	55	92	29	6
6	Kurubara Gollahalli	Agriculture land	2	9	1	
		Built-up	1	2		
		Commercial plantation	1	1		
		Others	1	1		
		Sub total	5	13	1	0

(continued)

Annex 2.7 (continued)

Sl. no.	Village name	Land use class	Bore well status		Open well status	
			Defunct	Functional	Defunct	Functional
7	Vadigenahalli	Agriculture land	1	6		
		Built-up	1	1		
		Others	2	2		
		Sub Total	4	9	0	0
		Grand Total	144	269	54	13

Annex 2.8 Bore well depth and land use relationship under Jadigenahalli gram panchayat

			Bore well depth range(in feet)							
Village name	BW status	Land use category	Depth not available	< 500	501–750	751–1000	1001–1250	1251–1500	> 1500	Total no of bore wells
Govindapura	Defunct	Agriculture land	3							3
		Built-up	2							2
		Commercial plantation	1							1
		Others	3							3
	Functional	Agriculture land		7	1	3				11
		Built-up						1		1
		Commercial plantation		3		1				4
		Others		1		2				3
		Water bodies				1				1
		Sub total	9	11	1	7	0	1	0	29
Haraluru	Defunct	Agriculture land	1			6	1	1		9
		Commercial plantation	1			4	1			6
	Functional	Agriculture land		1		2	5	5		13
		Built-up		1						1
		Commercial plantation				2	2	1		5
		Others			1	1		1		3
		Water bodies			1					1
		Sub total	2	2	2	15	9	8	0	38

Jadigenahalli	Defunct	Agriculture land	17			3				20
		Built-up	1							1
		Commercial plantation	5							5
		New layout	2							2
		others	3							3
	Functional	Agriculture land		1	12	26	12	3		54
		Built-up				2	1			3
		Commercial plantation			1	5	2			8
		New layout					1	1		2
		Others			2	3	6	2		13
		Water bodies			2	1	1			4
		Sub total	28	1	17	40	23	6	0	115

Annex 2.9 Distribution and status of borewells under Jadigenahalli Gram Panchayat

Village name	BW status	Land use category	Bore well depth range(in feet)							
			Depth not available	< 500	501 to 750	751 to 1000	1001 to 1250	1251 to 1500	> 1500	Total No of bore wells
Karibeerana hosahalli	Defunct	Agriculture land	16							16
		Built-up	1							1
		Commercial plantation	3							3
		Others	4							4
		Water bodies	1							1
	Functional	Agriculture land				17	1			18
		Commercial plantation				4	2			6
		Others				2		1		3
		Water bodies				1				1
		Sub total	25	0	0	24	3	1	0	53
Kolathur	Defunct	Agriculture land	20	1		1				22
		Built-up	7							7
		Commercial plantation	9			4		1		14
		New layout	1							1
		Others	4			2	1			7
		Water bodies	4							4
	Functional	Agriculture land		5	6	3	25	16	1	56
		Built-up					3	6		9
		Commercial plantation				1	7	2	1	11
		New layout					1	1		2
		Others		2			6	1		9
		Water bodies					2	3		5
		Agriculture land								
		Sub total	45	8	6	11	45	30	2	147

(continued)

Annex 2.9 (continued)

Village name	BW status	Land use category	Bore well depth range(in feet)							
			Depth not available	< 500	501 to 750	751 to 1000	1001 to 1250	1251 to 1500	> 1500	Total No of bore wells
Kurubara Gollahalli	Defunct	Agriculture land		1			1			2
		Built-up				1				1
		Commercial plantation				1				1
		Others		1						1
	Functional	Agriculture land				5	2	2		9
		Built-up			1			1		2
		Commercial plantation				1				1
		Others				1				1
		Sub total		2	1	9	3	3	0	18
Vadigenahalli	Defunct	Agriculture land	1							1
		Built-up	1							1
		Others	2							2
	Functional	Agriculture land			1	1	2	2		6
		Built-up						1		1
		Others					2			2
		Sub total	4	0	1	1	4	3	0	13
		Total	113	24	28	107	87	52	2	413

Annex 2.10 Bore well depth and land use relationship under Jadigenahalli gram panchayat

Village name	OW status	Land use category	Depth of wells (in feet)												Total No of Open Well
			10	15	20	25	30	35	40	45	50	55	60	>60	
Govindapura	Defunct	Agriculture land						1							1
		Commercial plantation								1					1
	Functional	Agriculture land						1							1
Haralur	Defunct	Agriculture land						1	1	1	2		1		6
		Commercial plantation					1		1		3		2		7
	Functional	Agriculture land							1		2				3
		Built-up									1				1
		Commercial plantation									1				1
Jadigenahalli	Defunct	Agriculture land				1	4		1	1					7
		Commercial plantation									1				1
Karibeerana haosahalli	Defunct	Agriculture land	1												1
	Functional	Agriculture land											1		1
Kolathur	Defunct	Agriculture land		1	1	2	3	3	1	3			1		15
		Commercial plantation				2	1	1		4	2	1	2		13
		Others											1		1
	Functional	Agriculture land				2			1	1				1	5
		Commercial plantation						1							1
Kurubara Gollahalli	Defunct	Agriculture land									1				1
Total			1	1	1	7	9	8	6	11	13	1	8	1	67

References

Anonymous: Government of Karnataka Notification No.FEE 37 FDP 2017 dt.23.03.2017

Bengaluru Metropolitan Region (Revised Structure Plan –2031)

Bengaluru Mirror Bureau (Suchith Kidyoor, March 24, 2015)

Central Ground Water Board (2013): Ground Water Information Booklet, Bengaluru Rural District, Karnataka 2013

CGWB, Groundwater Information, Medak District, Andhra Pradesh. Faridibad, India: Central Groundwater Board (2007). Available at: http://cgwb.gov.in/documents/SR/Medak.pdf. Cited 8 April 2009

I.R. Calder, P.T.W. Rosier, K.T. Prasanna, S. Parameswarappa, Eucalyptus water use greater than rainfall input – a possible explanation from southern India. Hydrol. Earth Syst. Sci. **1**(2), 249–256 (1997)

N.R. Gangadharappa, M. Shivamurthy, S. Ganesamoorthi, *Diffusion of Eucalyptus Tree Species in Karnataka*, India. (2003). http://www.fao.org/docrep/ARTICLE/WFC/XII/0054-B1.HTM

M. George, 1986. Nutrients uptake and cycling in a young Eucalyptus hybrid plantation. My Forest, March 1986

V.R. Hegde, V. Shreedhara, V.S. Hegde, Changing land use/land cover pattern in the Kali River Basin in Western Ghats, South India. Curr. Sci. **66**(2) (1994)

M. Reddy, B.C. Prabhakar, K. V. Brungesh and M. R. Aashaadha: Trends of observation on wells, Remedial measures and feasibility studies of Hoskote taluk, Bengaluru rural district, Karnataka, India: International Journal of Combined Research & Development (IJCRD) eISSN:2321-225X; pISSN:2321–2241 Volume: 5; Issue: 1; January –2016

M. Joshi and K. Palanisami: Impact of Eucalyptus Plantations on Ground Water Availability in South Karnataka, ICID 21st International Congress on Irrigation and Drainage, 15–23 Oct 2011, Tehran, Iran

S. Manasi, K.V. Raju, S. Poornima, K.P. Rashmi, B.R. Hemalatha Climate Change at Peri-Urban Contexts A case study of Jadigenahalli Gram Panchayat. Paper in *Climate Change Challenge (3C) and Social-Economic-Ecological Interface Building*, ed. by Sunil Nautiyal et al (New Delhi, Springer, 2016)

NRSC, Land Use/Land Cover database on 1:50,000 scale, Natural Resources Census Project, LUCMD, LRUMG, RSAA, National Remote Sensing Centre, ISRO, Hyderabad Master Plan Hosakote (2014)

Organization for Economic Cooperation and Development (OECD): OECD Environmental Indicators (2003)

Priyadarshini Sen: Shaping the fringes for the expansion of Cities-A Study on Hosakote in the Greater Bengaluru Urban Area, Transactions I Vol. 38, No. 1, (2016)

P. K. Srimani, N. Prasad: Land Use and Land Cover Mapping by using Remote Sensing and GIS Techniques – a Case Study of Kasaba Hobli, Hosakote Taluk, Bengaluru Rural District, Karnataka, India. International Journal of Engineering Research & Technology (IJERT), Vol. 2 Issue 1, January- 2013 ISSN: 2278–0181

T. Shah, A.D. Roy, A.S. Qureshi, J. Wang, Sustaining Asia's groundwater boom: An overview of issues and evidence. Nat. Res. Forum **27**, 103–141 (2003)

World Bank (2010) Sustaining Water for all in a Changing Climate, World Bank Group Implementation Progress Report of the Water Resources Sector Strategy, The International Bank for Reconstruction and Development/The World Bank, Washington

P. Wyrwoll, India's groundwater crisis, GWF Discussion Paper 1228, Global Water Forum, Canberra, Australia (2012)

Chapter 3
Holistic Health for Well-Being

Abstract With climate change getting serious, projections indicate impacts on health as obvious. Global call for addressing health concerns are in the fore. Besides, given a bigger challenge of providing rural health care in India with costs and infrastructure challenges on the rise, traditional medicine is seen as an alternative option for improving the health status of the masses. This chapter narrates a public-private partnership model adopted by the Soukya Foundation under the AYUSH programme, is engaged in reviving the traditional health care systems. The chapter analyses various activities and their status; the approach adopted in implementation; strengths and challenges of Soukya Foundation approach; key findings based on a field survey (102 households and 10 focus group discussions) and perceptions of people who have availed treatment at AYUSH Gram DMRC, SOUKYA Foundation, and also offers suggestions towards making it sustainable.

Keywords Holistic health · AYUSH · Soukya foundation · PPP model

3.1 Background

Some of the obvious and direct repercussions of climate change disruptions relate to human health; as in the forthcoming decades, they could put the lives and well-being of billions of people at increased risk (Costello et al. 2009). IPCC states that 'climate is projected to increase threats to human health'. Climate change is expected to affect human health through various ways such as heat waves, floods, storms besides complex ways like altered infectious disease patterns, disruptions of agricultural and other supportive ecosystems, and potential population displacement, conflicts over depleting resources like water, land, fisheries (Pachauri et al. 2007). With worsening of climate, there could be more malaria, asthma impacting pollutants, fatal heat waves and poor water quality. Potential threats to public health are one of the many reasons that we need to act on climate. Much has been written and discussed regarding health, a recent study indicates that 22% of the global burden of diseases in economically poor countries is related to environmental factors

S. Manasi, K. V. Raju, *Coping Mechanisms for Climate Change in Peri-Urban Areas*, https://doi.org/10.1007/978-3-030-18517-6_3

(Pruss Ustum et al. 2017). Future impacts will also depend on the location of persons, their sensitivity to health threats, their exposure to climate change impacts and the ability of their community to adapt to change. The extent and intensity of these health risks depend on the ability of public health and safety systems as part of addressing or preparing for these changing threats besides other factors like individuals behavior, age, gender, and economic status.

The targets under the UN Sustainable Development Goal-3 indicate at ensuring healthy lives and promoting an overall well-being for all, at all ages, by fulfilling the right to respect and dignity and combating climate change uncertainties. The World Health Organization (WHO) has declared that climate disruptions can pose 'unacceptable risks' to global public health, while urging strong action to address climate change. United Nationals Conference on Climate Change draft resolution – an outline for the international climate agreement in Paris – focused on universal right to health and sustainable development (Pruss Ustum et al. 2017).

More specifically, in the context of health discussions, inter relations of human health and well-being with that of animal health and a healthy environment have been gaining more attention in the last two decades. Varied policy perspectives have emerged and communities of practice in global health, where it is termed as 'Environment and Health', 'Planetary Health', 'Biodiversity and Health', 'Eco-social Health', 'Climate Change and Health', 'Conservation Health' and so on. 71st World Health Assembly was conducted wherein they deliberated on the broad environmental and social drivers and emerging perspectives on health (WHO 2018). The State of the Knowledge Review on Biodiversity and Health published in 2015 by the Secretariat of the Convention on Biological Diversity (SCBD) and the World Health Organization (WHO) while assessing the multiple linkages, highlights the nexus between Biodiversity and Health through various thematic areas, including traditional medicine and natural products of health and nutrition (WHO 2015). Inter-linkages between human, animal and environmental health are better appreciated with the emergence of 'One Health'.

This has close parallels to the 'Eco-Health' approach where 'One Health' focuses on diseases at the human-animal interface or food borne diseases, Eco-health also focuses on broader issues like climate change and ecosystem links. These new perspectives go beyond traditional notions of access to health care to reasserting the WHO definition of health as 'State of complete physical, mental and social well-being not merely the absence of disease or infirmity. They have resulted in a better understanding of the determinants and drivers of sustainable good health and well-being with their critical linkages opening new spaces for education, research and development and international co-operation. They have also flagged methodological opportunities of systems thinking, transdisciplinarity, problem based social learning, among others (Payyappallimana 2018).

Given this backdrop, the significance of indigenous/traditional healing systems is gaining prominence. The global focus seems to be shifting from 'alternative medicine' to 'integrative health'. The holistic wisdom of Ayurveda, Yoga and other ancient health systems offer the necessary wisdom, experience and capabilities that are crucial to such transformational change. Innovative approaches where Indian health systems and modern medicine synergized together can realize the dream of futuristic integrative health systems (Patwardhan 2018).

Ancient healing traditions prevailing as far back as 5000 years ago in India and China, give importance to living a healthy way of life in harmony with nature. Holistic medicine is an ancient system of health care that dates back to thousands of years ago, even though the expression is of a much more recent era. Although the term *holism*[1] was introduced by Jan Christiaan Smuts in 1926 as a way of viewing living things as "entities greater than and different from the sum of their parts," it got more prominence only from the 1970s onwards (Wang Chunxin 2017). Holistic health[2] approaches and healing are practised in several countries. Recent research has discovered that ancient Chinese herbal remedy is effective in dealing with chronic pain. Traditional Chinese medicine (TCM) is holistic in its approach. Treating patients holistically was also well understood by Greeks. In Ancient Greece, patients would visit Asklepion, holy temples of healing, where diseases were cured by harmonizing all the social, environmental, psychological, spiritual and physical factors that were believed to interact in the causation of illness. The early Mayan civilizations also had developed healing systems that were holistic in nature. Mayans believed that health was all about 'balance'. Man was considered as an integral and interactive part of the cosmos and society and any imbalance would lead to illness. To cure imbalances, the Mayans used a holistic approach to healing that focused on both the spiritual and physical aspects of well-being. In Chile, the Ministry of Health hopes resurgence of traditional medicine. In African traditional medicine, the curative, training promotive and rehabilitative services are referred to as clinical practices. These traditional health care services have been based in through tradition and culture (Mhame et al. 2001–2010).

Given this backdrop, in India, the Ministry of AYUSH represents a pluralistic and integrative scheme of health services (AYUSH- Ayurveda, Yoga, Naturopathy, Unani, Siddha, Homeopathy). AYUSH is witnessing a highly receptive environment with the value of AYUSH systems in health care being widely recognized. Many reforms in administration, research strategies, nature of practice and education are already underway at the Ministry of AYUSH.

The scope of NHP 2017 in the context of AYUSH is extensive. It generally recognizes medical pluralism as reality in India with an amplified role for AYUSH (Shankar and Patwardhan 2017).

---

[1] The American Holistic Health Association (AHHA) describes holistic health as 'rather than focusing on illness or specific parts of the body, this ancient approach to health considers the whole person and how he or she interacts with his or her environment. It emphasizes the connection of mind, body and spirit. The goal is to achieve maximum well-being, where everything is functioning the very best that is possible'. Mind, body and spirit – these aspects of each patient are considered to be intertwined and essential to achieving optimal health. Human beings are comprised of physical, emotional, mental and spiritual elements is another way of looking at it. It focuses on natural healing.

[2] While holistic medicine may incorporate treatments from mainstream health care, drugs and surgery are not the line of first defence. By definition, holistic medicine is about lifestyle changes, non-invasive remedies and improving the flow of a person's life-force energy, along with enhancing the body's own ability to heal itself (Wang Chunxin 2017).

The Ministry of AYUSH, Central Council of Research in Ayurveda and Siddha, and numerous other collaborative centres of WHO in India are assigned with several Appraisal Project Work (APW) and Direct Financial Cooperation (DFC) projects aimed at strengthening Ayurveda, Homeopathy, Naturopathy and Yoga as evidence-based systems of medicine for their global acceptance. Implementation of pharma co-vigilance programmes in Ayurveda and Homeopathy, publication of documents for rational use and initiatives to prepare consumer guidelines for an appropriate use of Ayurvedic medicines are some other contributions of WHO towards the advancement of Ayurveda at national as well as global levels (Nancy 2011).

With this detailed background on climate change, health concerns and the role of holistic health, we now proceed to discuss these concerns in peri-urban contexts, where food and nutritional security have become more critical as people living there are mostly socio-economically marginalized. Peri-urban farmers have shifted to the cultivation of growing vegetables, mostly to cater to the urban markets. It is interesting to note that, in urban contexts, realizing the importance of chemicals-free food, 'for-natural food' initiatives have evolved that need mention in this context of peri-urban Bengaluru with urban home gardens gaining prominence around Bengaluru city. Agriculture Man Ecology (AME), is promoting organic terrace gardening in Magadi village near Bengaluru with an emphasis on sustainable and healthy cities. Sukrishi Farm, an organic farm near Nelamangala is another example of nature-friendly living that promotes organic farming, and creates awareness among the urbanites and provides them with an experience of serene living. All these initiatives are focused on improving the quality of life and well-being of people at large.

Thus, peri-urban environments play an important role in shaping human health and well-being. Considering that a holistic approach is essential to addressing health issues pertaining to these environs, AYUSH initiative – a new program, with its implications – is analyzed in this chapter. We also discuss several important aspects that have led to its success and adaptability besides offering few suggestions for its sustainability for environmental well-being as well. Given that, India has a rich heritage of traditional medicine with the traditional health care systems flourishing through many centuries with holistic health care being always the norm. Hence, medicinal plants-based traditional systems have always played an important role in providing inclusive health care to a large section of population (Ravishanakar and Shukla 2007). The Indian health system has six recognized systems of medicine in this group- Ayurveda, Siddha, Unani, Yoga, Naturopathy and Homoeopathy (Rajeev 2012). World Health Organization has also evinced interest in the traditional systems of medicine with a mandate from the United Nations Organization to promote health programmes of every member country.

AYUSH project in Karnataka is a Government-sponsored programme under public- private partnership (PPP) mode for a period of 3 years, with the main aim of providing a complete health care and is designed to generate cost-effective and affordable health care systems. In brief, the programme focuses on "*Preventive, Curative and Promotion*". The overall health check up of the entire population is done by AYUSH doctors based on AYUSH systems and are provided with basic knowledge for promotion of health and prevention of diseases. The communities are educated in healthy practices and advantages of traditional food items used locally and their medicinal properties. AYUSH doctors also undertake health check-up camps at schools in and around the selected villages. AYUSH training is also imparted to ASHA workers, Anganwadi workers, school teachers etc.; awareness building activities are also conducted through gram panchayats involving schools, anganwadis, self-help groups and other community organizations.

As part of reaching out to more people, they involve Accredited Social Health Activist (ASHA) workers, Anganwadi[3] workers, and school teachers with a focus on the basics of health, hygiene, local health traditions, using herbs, plants and traditional food practices. There is a good rapport with the medical officer of the Primary Care centre. The programme is designed to witness: (a) Significant improvements in the prevailing disease conditions and preventive medicine; (b) Considerable awareness regarding AYUSH systems of medicine among people; (c) Yoga as a norm of life; (d) Improved Quality of life; (e) Every household with a few medicinal plants in its home garden; (f) First line of treatment is understood and practised by all.

There are specific indicators used for marking the progress and achievements – (a) Analysis of health records; (b) Investigations; (c) Interactions with the public; (d) Physical verification of cultivations. Under AYUSH Grama project, a total of 10 Grama panchayats of nine districts were selected in Karnataka state, by the government of Karnataka. These districts were handled by 10 different Non-Governmental organizations (Annex 3.1). All these organisations were known for addressing health issues, using traditional methods in their own unique ways and approach. This study specifically focused on Dr. Mathai's Rural Centre (DMRC), Soukya Foundation intervention in improving community health care and the process adopted are described[4].

[3] Anganwadi is a type of rural child care centre in India. They were started by the Indian government in 1975 as part of the Integrated Child Development Services program to combat child hunger and malnutrition. Anganwadi means "courtyard shelter" in Indian languages.

[4] ***Organisational setup and Implementation process***

DMRC SOUKYA Foundation has a meticulous implementation process with 11 staff members and 4 ASHA workers and is headed by Dr. Issac Mathai, Managing Trustee and Mrs. Suja Issac, Trustee of SOUKYA Foundation Charitable Trust. The monitoring and supervising are taken care of by Dr. Sudha, Coordinator DMRC, SOUKYA Foundation, while Ayush Grama, Jadigenahalli, is handled by Dr. Satheesh, Dr. Manjunath Dr. Ganesh, and Dr. Prakash. Dr. Ganesh, apart from managing his responsibilities as a doctor in treating patients, handles other responsibilities besides sending daily reports to Dr. Issac Mathai, Mrs. Suja Issac and Dr. Sudha regarding patient details, activities and meetings conducted and other activities. Administration and accounts are handled by Accounts Department of DMRC, SOUKYA Foundation. Government co-ordinations are handled by staff of SOUKYA Foundation registered office at Richmond circle, Bangalore. Pharmacist holds specific responsibilities - maintains patient files, distributes medicines, prepares purchase order for stock procurement, update records and provides medicines prescribed by doctors. Asha workers also help in the hospital activities on a daily basis. Weekly meetings are held every Wednesday at DMRC, SOUKYA Foundation, with discussions on work processes, changes required and plan of action. The feedback and discussions are taken into consideration and changes incorporated into the programmes.

3.2 Dr. Mathai's Rural Health Centre (DMRC) Soukya Foundation[5]

We focused on understanding the challenges in the process of improving public health through traditional systems of medicine, in the context of allopathic system of health care being popular and also to comprehend the dynamics of health care from a socio-economic and environmental perspective. As part of methodology, stratified random and purposive sampling methods were adopted for capturing the perceptions of patients based on their regular visits and follow-up of treatment, supported by focused group discussions. Primary data was collected from households using a structured questionnaire, covering socio-economic aspects, AYUSH activities, preferences and perceptions of health care centres, access to health care, sources of water, sanitation and hygiene, cropping patterns, food habits, user satisfaction, religious beliefs, medicinal plants cultivation and usage, ecology and cultural aspects as part of receiving a systematic public feedback in terms of implications and perceptions. Responses were collected from a total of 102 households (10% of the sample).

At the empirical level, the study used data sets from AYUSH Grama DMRC, SOUKYA Foundation. Based on the secondary data and discussions with doctors at DMRC, SOUKYA Foundation, a pilot study was done and research tools developed. It was further segregated based on the number of follow-ups by patients; 70% of the total sample was drawn from >10, 5–10 and 2–5 follow-ups respectively and 30% from one follow-up and one-time visit. The samples were drawn

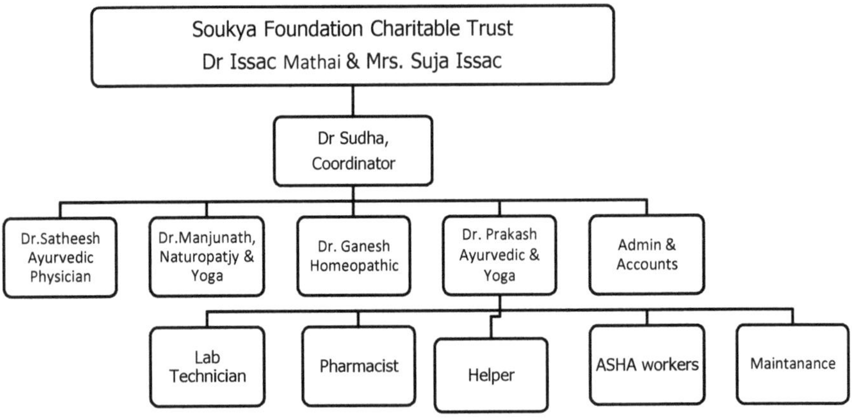

Source: Authors, 2014

[5] SOUKYA Foundation Charitable Trust was established in Bangalore by Dr. Issac Mathai in 1996. the Government of Karnataka, with the support of Department of AYUSH, Department of Health and Family Welfare and SOUKYA foundation, a charitable trust based in Bangalore, planned to establish a Rural Health Care Centre in Jadigenahalli, Hoskote Taluk, Bangalore Rural district.

from landless labourers and small, marginal and large farmers to make it further representative.

The SOUKYA Foundation[6] has adopted both holistic and integrated approach for treating patients. The patients are treated based on their response to the treatment. Patients are exposed to either only (a) Homeopathy; (b) Homeopathy and Ayurveda; (c) Homeopathy, Ayurveda and Yoga; (d) Homeopathy, Ayurveda, Naturopathy, Yoga. There are 22 activities as per the design of AYUSH Grama, out of which, 9 activities have been already initiated and are found functioning satisfactorily, while 7 are in the initial stages and 5 are yet to be initiated; one activity though has proved less successful. Table 3.1 presents the status of DMRC, SOUKYA Foundation activities under Jadigenahalli Grama Panchayat[7,8].

3.3 Key Findings

3.3.1 Awareness Creation

DMRC, Soukya Foundation has adopted different approaches towards creating awareness regarding its activities and their purpose (Fig. 3.1). To build confidence among the people about the AYUSH approach, they make announcements,

[6] Partial funding is provided by Karnataka Government for a period of 1 year 3 months including the establishment cost of Rs 34.45 lakh which is reimbursed after the programme, while the rest of the expenditure is borne by Soukya Foundation. In addition, SOUKYA Foundation, in Memory of Rev. Fr. Mathai Nooranal, has donated a building worth Rs 45 lakh for this purpose and the space for setting up a Health Care Centre is provided by the Government of Karnataka.The Government has also been supportive in providing educational and awareness creation materials - posters and booklets.

[7] Jadigenahalli Grama Panchayat has 3 Private medical clinics and 1 Primary Health Centre (PHC). Patients generally visit one of these clinics for treatment. The PHC largely caters to Gynaecology related concerns and hence, the private clinics are preferred to avail of other treatments. Patients from low income groups largely restrict themselves to private clinics, PHC, and government hospitals located in Hoskote or Malur (12 Km Distance), while, patients belonging to high income groups travel to private hospitals in Hoskote as the next best option in terms of proximity and accessibility. Government of Karnataka, with the support of Department of AYUSH, Department of Health and Family Welfare and SOUKYA foundation, a charitable trust based in Bangalore, planned to establish a *Rural Health Care Centre* in Jadigenahalli, Hoskote Taluk, Bangalore Rural district. This centre was the first of its kind in *Rural India* with different systems of Medicine like Ayurveda, Unani, Siddha and Homeopathy integrated. The AYUSH Grama DMRC, SOUKYA Foundation Started the work process from 2012. SOUKYA Foundation was one among the ten organizations selected under the State Government's Programme AYUSH. With DMRC, SOUKYA Foundation being close to the PHC, several patients tend to visit it to check an alternative option.

[8] • ***Mother and Child Care / ANC & PNC Care***

This programme mainly Totally, 2 programmes are conducted for Anti Natal Checkup (ANC) and 27 women were treated since March 2013 (see Annex). The Anti Natal Checkup has been started in Ayush Grama DMRC, SOUKYA Foundation, from March 13th 2014. Till now, two checkups have been conducted with a total of 11 women tested and given Ayurveda medicines (Sukaprasawa: 2 tea spoon early in the morning on empty stomach and suggested to use warm water after taking medicines). In order to make them develop a habit of doing these exercises, doctors provide awareness about the frequency of C-section surgery due to irregular physical exercises.

Table 3.1 Activities and status. (Source: Soukya Foundation Documents)

Sl. No	Activities initiated	Initiated and successfully ongoing	Not yet initiated	Less successful
1	**Health care activity**			
	a. Mental health programme			
	b.Panchakarma[a]			
	c. Ksharasutra[b]			
	d. Balopachara programme[c]			
	e. Mother and child care[d]			
	f. Prevention of viral diseases[e]			
	g. De-Addiction promotion programme[f]			
	h. Geriatric care[g]			
	i.Anaemia programme			
	j.Diabetic care			
	i. Home remedies programme[h]			
	j. Home medicinal kit distribution[i]			
2	**Nutrition programme**[j]			
3	**Educational activities**			
	a. School health(School yoga, ARSH programme for high school children)[k]			
	b. Survey of RET (Dept. of school herbal garden)[l]			
4	**Livelihood components**			
	a. Encouragement of local health traditions			
	b. Job oriented training for villagers			
	c. Providing marketing facilities			
4	**Ecological and environmental conservation and restoration**			
	a. Construction of smokeless chulas			
	b. Goshala Nirmana (Cow shelters)			
	c. Rain water harvesting			
	d. Solid waste management			
	e. Cultivation of medicinal plants			
	f. Herbal and kitchen garden			
5	**Agriculture/organic farming**			
6	**Technology up gradation/capacity building/Training**			

[a]Panchakarma relates to cleansing of the body according to Ayurveda. a therapeutic way of eliminating toxins from the body.
[b]This is a treatment for patients suffering from Piles and Fistula. ***Panchakarma Therapy*** - These treatments are chargeable; however, the cost of treatment provided is generally 20% less than the normal price charged.
[c]Treatment is provided for children below 12 years through homeopathic medicine and aims to boost immunity. ***Balopachara (Paediatric Care)*** - Totally, 586 children have been treated so far under Balopachara Treatment. A research study is conducted to document the responses of children to treatment. After a medical checkup, children with ailments are given homoepathy/ayurveda medicines and are kept under observation to check their response to medicine/ variations/improvement. Nutritional supplement has been initiated recently. Swarna Bindu treatment is yet to begin (see Annex).
[d]Focuses is on maternal health and neonatal care of infants. Medical kit is provided and across the stages of semester, Ayurvedic medicines are provided and explained to couples. Expectant mothers are given Ayurveda medicines. Expectant mothers are checked for physiological changes besides, encouraging them on life style changes - food habits. music for mental health, *Pranayama* (Breathing exercises) and *sukshayama* (Finger loosening, eye and head exercise) to promote normal delivery.
[e]Prevention of diseases through homoeopathic preventive medicine for ailments like Flue, Dengue, Chikungunya
[f]Addiction is defined as a chronic relapsing disorder for people using substances like smoking, alcohol, rave drugs, and medical drugs. Ayush Grama is largely able to influence the people through promotion of De-addiction programmes by creating awareness and providing homeopathic medicines.
[g]Geriatric Care is all about providing treatment to the elderly population beyond 60 years of age

Table 3.1 (continued)

[h]Home remedy treatments are aimed at curing diseases or ailments with commonly available spices, vegetables, herbs or other plants.

[i]The medical kit is a handy pouch containing homeopathy medicines for 28 different kinds of diseases (See Annexure) and is distributed free of cost.

[j]Nutrimix is prepared by using various cereals, pulses and millets,given to malnourished people for consumtion daily once a day.

[k]Yoga Therapy programmes are conducted regularly across all the villages at various locations to spread the message of Yoga in curtailing ailments.

[l]Soukya identified 15 endangered species which are cultivated near its health centre, in a demo farm.

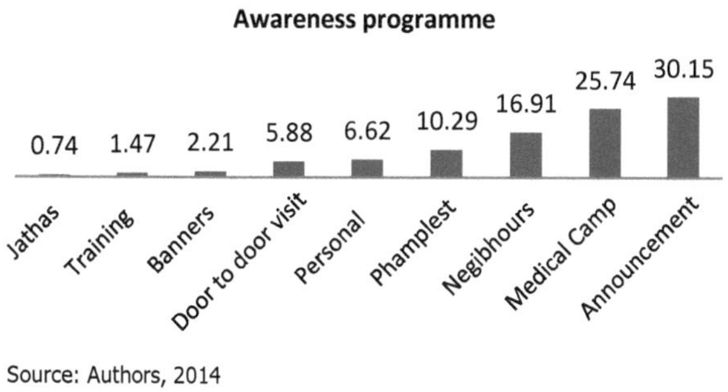

Fig. 3.1 Awareness creation and popularity (%)

- ***Prevention of Viral Diseases***

AYUSH approach towards prevention of viral diseases is holistic in that it works on, etc. Trainings are conducted for creating awareness about cleanliness, personal hygiene through daily bath, washing hands before food and need for using toilets. Totally, 124 trainings have been conducted so far. Besides this, training is also provided in the preparation of natural mosquito repellents from locally available plants such as Neem and marigold plants. About 190 homes covering 760 people have been given preventive Homeopathy medicine.

- ***Vector Borne Disease***

This programme aims to create awareness about vectors. A total of 3 training programmes have been conducted in Jadigenahalli, Vadigenahalli and K. B. Hosahalli. On an average, 30 -40 people attended the programme. The programme covered awareness about (i) Type of vector borne diseases; (ii) Incubation period; (iii) Manifestation of diseases; (iv) Complication of disease; (v) Measures to control the disease; (vi) Cultivation of medicinal plants like Marigold, Eucalyptus, Tulasi etc around houses to prevent mosquitoes as fragrance of these plants helps reduce mosquitoes; (vii) Training is given to prepare homemade repellents using home ingredients like Bajee, Vidanga, camphor, sambharani powder, cow dung (dried and fine powder), gomuthra, arka etc; all these have to be mixed and made into a 5cm long stick that can last for 35 min.

- ***Medicinal Kit***

Medical kits that can cure minor ailments costing medicines worth Rs.300 were provided to 200 families.

- ***De-Addiction Promotion Programs***

A good number of patients have been treated in DMRC, SOUKYA Foundation, through Yoga and Homeopathy medicines. The addicted people are identified during personal visits by Doctors and ASHA workers. Alcoholic patients who take the initiative and visit DMRC, SOUKYA Foundation, are treated in the hospital while some unwilling patients are treated by doctors through personal visits and persuasion.

- ***Geriatric Care***

Special Medical Camps are conducted to treat the elderly. This programme has been recently introduced with the patients treated so far numbering 393.

distribute pamphlets and banners. Besides this, they conduct free medical camps (every month) where they screen slides for patients about diabetes, hypertension and other ailments, besides prescribing medication as well. Announcements and medical camps have had a huge impact in terms of creating awareness. Besides, awareness creation has helped spread positive effects in terms of people, who have availed themselves of treatment at DMRC, SOUKYA Foundation, informing neighbors and sharing their experiences. This has indirectly promoted the programme with a considerable impact on popularising the AYUSH approach.

This programme has successfully reached out to all groups of people with more than 500 patients treated. So far, they have conducted 12 village health education programmes involving 409 participants; Vector Borne Diseases programme has covered 124 households; Home medicine distribution drive has covered 48 households.

3.3.1.1 Promotion of AYUSH Through ASHA Workers[9]

DMRC, SOUKYA Foundation has trained and engaged ASHA workers as catalysts in promoting various programmes and encouraging people to avail benefits. The programme has evovled and needs streamlining with the involvement of ASHA

Goshala Nirmana was initiated with people actively participating and Goshala Training but was dropped as it had not been included in the action plan of the Department. Further, people wanted funding for establishing Goshalas.

People attending yoga classes

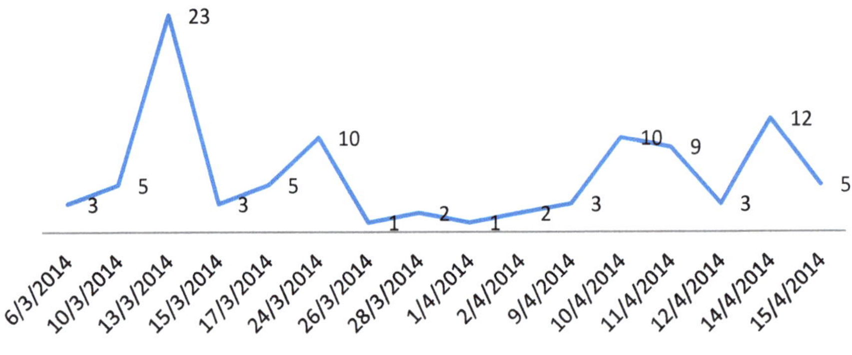

Source: , AYUSH Grama DMRC, SOUKYA Foundation, 2014

[9]Role of ASHA workers – Motivating patients to visit DMRC, SOUKYA Foundation, through door-to-door visit every day, Identify persons in need of health care, advising them to visit DMRC, SOUKYA Foundation, for treatment, Conducting patient Survey and reporting the common diseases observed across the villages to DMRC, SOUKYA Foundation doctors, Collect details of patients – name, age, sex, disease type, with signature entered in their daily register before reporting to the Doctors once in a week, recommending home remedies for minor diseases, promoting health and hygiene, pooling villagers for medical camps and other activities conducted by Soukya. Participating in camps and supporting doctors in conducting camps, attending training programmes and meetings conducted by Soukya.

workers: they have been trained in the use of medicinal plants, health and hygiene, vector borne diseases, and disease identification.

3.3.1.2 Awareness Creation Regarding Medicinal Plants

Survey of Rare and Endemic Trees (RET) and greater preference is given to growing medicinal plants locally and organically. This has been with the hope that it could promote livelihoods. But cultivation of medicinal plants has not evinced much interest due to water scarcity and difficulties involved in adapting to new cultivation options. Promotion of home remedies programme in educating people to prevent minor illnesses at affordable cost and aid reduced medical costs has found a positive response with supportive training programmes and imparting knowledge by ASHA workers. Herbal and kitchen gardens have been well received with a good participation rate in training programmes.

Interestingly, across all schools, teachers have helped children understand the medicinal value of locally available common plants. In KB Hosahalli village school premises, teachers have involved children in the cultivation of medicinal plants. With the intervention of DMRC, SOUKYA Foundation, these activities have got a good boost. In Jadigenahalli, 200 plants had been procured from the forest department for schools; however, being faced with monkey and goat menace they are unable to protect them during holidays. However, it does indicate that both teachers and students have taken a keen interest in protecting plants.

3.3.1.3 Promoting Physical Fitness and Hygiene

Popular Yoga

Yoga is a welcome initiative bringing about considerable changes among the village folk. Children experiencing health benefits are practising it regularly. Improvements in children's health have also attracted adults and senior citizens towards practising yoga. Yoga classes are already part of the curriculum in schools, hence, is compulsory and practised even before SOUKYA Foundation's initiative. However, yoga classes conducted by DMRC, SOUKYA Foundation have reinforced and strengthened the programme. Discussions with teachers at school indicate that the teaching approach of DMRC, SOUKYA Foundation, is appealing with a focus on theory by explaining aasannas (postures) and their associated benefits has made it interesting for children. Pranayama and Surya Namaskhar are taught as part of improving children's fitness, health and stamina, especially their concentration levels (Plate 3.1). Apart from having a positive impact on children, (physical fitness and flexibility), the yoga programme has succeeded in influencing even parents to learn yoga. Besides, yoga has influenced the lifestyles and food habits of children. With women showing interest in learning yoga, separate sessions are being held for them.

Plate 3.1 Children practicing yoga in schools, Jadigenahalli village

Inculcating Sanitation Practices

Health and hygiene practice among children by encouraging them to keep their surroundings clean and maintain personal hygiene is regularly monitored. Monthly checkups at Anganwadis[10] for malnutrition, physical growth (height, weight), hemoglobin levels and respiratory ailments have aided in reducing morbidity. Children diagnosed with ailments are treated with ayurveda/homeopathy medication and monitored. A research study to document children's responses to medication and treatment provided for low hemoglobin count has been taken up. Totally 163 children have been provided with treatment so far for various ailments and also 40 training programmes have been held for Anganwadi teachers.

Handling Potable Water and Nutritional Supplements

Potable water scheme aims at creating awareness regarding protection and use of clean water and simple methods of purifying water for drinking/cooking (Plate 3.2). Nutrimix powder provided to patients has been welcomed by people.

[10]The Indian government started Anganwadi in 1975 as part of the Integrated Child Development Services programme to combat child hunger and malnutrition for children aged between 2 and 5 years. Several programmes are taken up under this scheme viz. education, nutrition supply, supplementation, pre-school activities

Plate 3.2 Eating healthy – promoting nutritious food habits among children by Ayush Grama Soukya Foundation

Box 3.1: Ayush Pushti Biscuits

To start with, the state Health department has come up with a nutritional solution- Ayush Pushti biscuits scheme to address the problem of malnutrition among rural children in the State. On a pilot basis, the scheme will be launched in Bijapur and Jamakhandi Taluks in North Karnataka. These are made using ayurvedic herbs (See Table), while taking into account the nutrition needs of anganwadi children. This scheme is intended to help improve digestion among children and to boost their immunity and also their concentration levels. Ingredients of Pushti biscuits: (listed them in continuation to save space with english equivalents provided)

Wheat	Amrutha balli
Ragi	Vidanga
Green gram	Vache
Groundnut	Kushmanda (pumpkin)
Ashwaghanda	Gokshura
Brahmi	Jestamadhu
Pippali	Ksheera
Sugar	Grutha (ghee)

3.3.2 *Experiences and Perceptions*

Since access to medical care under Jadigenahalli Grama Panchayat is restricted to only 3 Private medical clinics and one primary health care center (PHC) that largely caters to gynaecology-related concerns, private clinics are the only other option. Patients from low income groups largely restrict themselves to private clinics, PHC, and government hospitals located in Hoskote or Malur (12 Km away), while, patients belonging to high income groups travel to private hospitals in Hoskote as the next best option. With DMRC, SOUKYA Foundation being close to the PHC, several patients tend to visit it as a good alternative option. Respondents prefer private clinics, although they charge exorbitantly, expecting a better treatment. Hence, DMRC, SOUKYA Foundation's intervention is considered a welcome development, hoping for an improved health care. Women feel blessed with DMRC Soukya Foundation at their vicinity as it reduces their long hours of wait for transport to avail of treatment at Hoskote town.

For experiencing positive results of the treatment, 62% of the people consider patients' attitude, timely medication and follow-up visits as important. This requires time and effort on the part of patients which is lacking. These have been elicited from those visiting SOUKYA Foundation. They have also realized that homeopathy medicines take time for positive results besides requiring regular follow-up visits.

The number of patients visiting SOUKYA Foundation shows a mixed trend across months and villages (Fig. 3.2). A large number of people who visit health camps avail of the benefits associated with the programme. Hence, the initial months show a surge in patient numbers. However, the follow-up visits for medical care by patients depend on the health-related ailments – chronic or common ailments. People appreciate the positive approach of doctors, as they spend a considerable time listening to patients and recording details. Several patients have experienced improvements in

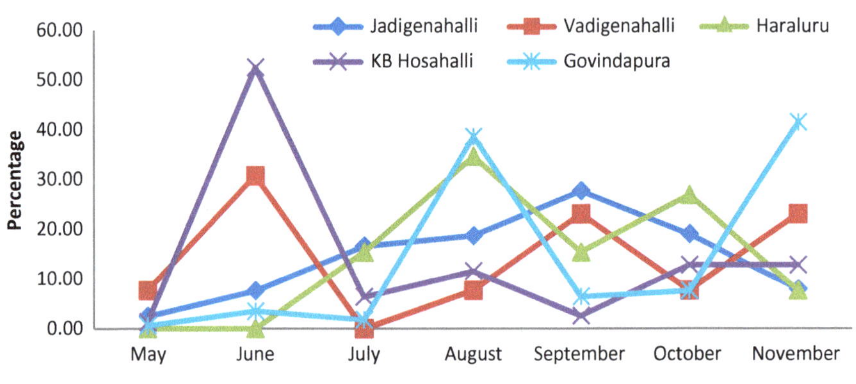

Fig. 3.2 New patients visiting DMRC, SOUKYA Foundation (May-Nov 2013)

their health status with changes in lifestyle combined with medication. 26% of the people have been visiting Soukya Foundation for the past 6 months, followed by 23% once in a month. Only 5% of the people are visiting this Foundation once a year. In a short time, DMRC, SOUKYA Foundation, has been able to draw attention of the villagers through its holistic and traditional approach of medicines and treatment. However, follow-up visits by patients continue to be an issue due to busy schedules and negligence.

A large number of people are disseminates through word of mouth to their own family members and relatives/friends to avail treatment at DMRC, SOUKYA Foundation, based on their own experiences. Overall, 66.66% of the people have influenced others, 41.17% have influenced their own family members and 15.68% have influenced their relatives to the extent of traveling from neighboring villages. On the performance of DMRC, SOUKYA Foundation 34% have rated as good, while 29% as excellent. The doctors' approach and interactions with patients have had positive impacts in terms of curing ailments and 17% feel that there is an improvement in medical equipments; Ayurveda and lady doctors adhere to women's health issues.

3.3.2.1 Treatment for Different Ailments

Most of the patients visiting DMRC, SOUKYA Foundation, generally complain of body pain (22.5%), followed by diabetes (16.7%), fever (11.8%) and gastricitis (11.8%) (Fig. 3.3). Doctors observe that more people in villages are tending towards diabetes, gastric and weakness problems owing to irregular food habits, junk food consumption, foods with least nutrition, preference for aerated and carbonated drinks. Changing lifestyles and an inadequate understanding of good health add to misery.

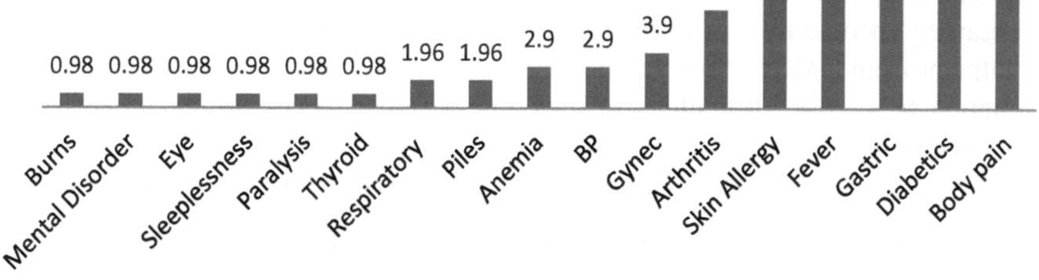

Source: Authors, 2014.

Fig. 3.3 Patients with different kinds of aliments (% to total)

3.3.2.2 Perceptions Regarding Treatment and Cure

Alternative medicines appear to have impressed patients, as they have experienced positive results with homeopathy treatment as compared to allopathy; 41% of patients got partially cured of their ailments and 32% fully (all ailments), There are patients (27%) who do not take medicines regularly, owing to wrong perceptions about taste and uncertainty about their effectiveness.

a. Priority for Diabetes and Hypertension
Diabetes and hypertension are the common ailments observed across all five villages. Earlier, 8.8% of BP and 13.73% of diabetic mellitus patients preferred Hoskote and private clinics under Jadigenahalli MP. However, after the intervention by Soukya, 5.88% BP and 11.76% of diabetic patients have chosen DMRC, Soukya Foundation for treatment. This could be attributed to the time spent with patients and awareness creation programmes. However, changes may be due to other factors also like a reduced monthly medical expenditure.

b. Treatment for Arthritis, Respiratory Disorders and Hyper Acidity
Treatment at DMRC, Soukya Foundation is largely preferred by arthritis and hyper acidity patients. Those on allopathy drugs have changed to homeopathy medication as they found homeopathy was comparatively more effective; 14.71% hyperacidity, 11.76% arthritis patients have also moved from allopathy to homeopathy. But, in the case of respiratory ailments, no changes have been observed. Patients experiencing positive health improvements have spread the message to friends and relatives, resulting in more number of patients from neighboring villages seeking homeopathy treatment.

c. Preference for Treatment of Fever, Body pain and Allergies
For common ailments like fever and body pain, patients prefer treatment in private clinics and hospitals at Hoskote. This could be due to a general perception that homeopathic and ayurvedic treatment takes a longer time for patients to be cured of their ailments. This prolonged time affects their daily routine and ability to work. Allopathy is seen as more effective in delivering quick results. However, with respect to body pain and allergies, people prefer DMRC, Soukya Foundation for treatment (24.51% patients for body pain and 8.82% for allergic patients) as they feel that allopathic medicines are not effective.

Largely, people from socially backward and economically weaker groups seek treatment from DMRC, Soukya Foundation because of reduced costs with no consultation fees. Also, difficulties in travelling to other places (Malur or Hoskote) could be reduced.

Besides, positive benefits with respect to cure and comfort with no side effects.

d. Perceptions about Ayurveda and Homeopathy Systems

At large, respondents are aware of ayurveda (70%), but did not take any ayurvedic medicine. With DMRC, Soukya Foundation creating awareness regarding ayurveda, respondents have started taking ayurvedic medicine (18%) and are convinced of its positive impacts. Respondents are being introduced to homeopathy with DMRC, Soukya Foundation's intervention and 48% of them report having experienced good results, rating the treatment as positive (Figs. 3.4 and 3.5). However, there is a misconception among some respondents (20%) that sugar coated tablets do not cure ailments and that they rather aggravate ailments.

Encouragement of Local Health Traditions promotes good traditions by documenting before encouraging the rightful traditions which can be approved scientifically. Traditions like praying, meditation, circumcision (Muslim tradition) are encouraged, while belief in ghosts and social drinking as a regular practice are subtly discouraged.

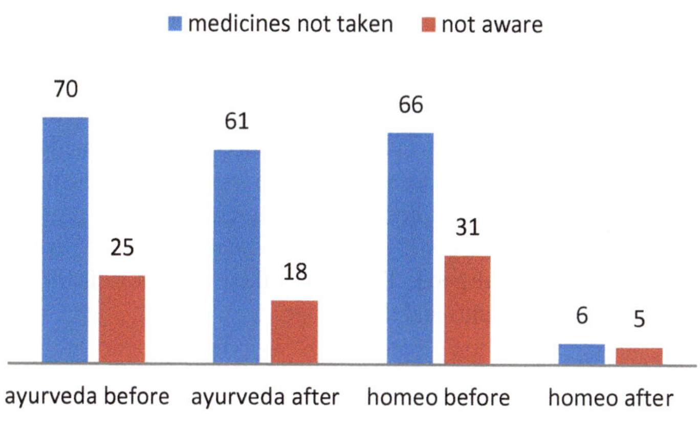

Source: Authors, 2014.

Fig. 3.4 Perceptions about ayurveda and homeopathy

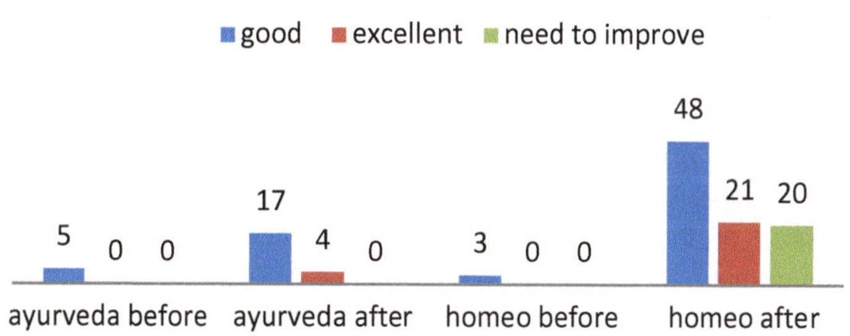

Source: Authors, 2014.

Fig. 3.5 Rating of ayurveda and homeopathy

3.3.2.3 Most Popular Programmes

A majority of the people are observed benefitting from DMRC, Soukya Foundation's activities. Some of the activities have created a unique identity for DMRC, Soukya Foundation, mainly because of its approach and treatment. However, with respect to the medical camps conducted, 51% attended with 45% exploring the benefits of diabetes and hypertension screening. A total of 44% of the people participated in yoga, with 26% benefitting through changes in lifestyle and food habits, while women groups opine that yoga programmes should be conducted regularly as they would help them remain disciplined and confident.

Home remedy programme is gaining popularity with the participation rate gradually increasing across all the villages. This programme has been able to draw 30% of the people's attention, with 24% of them benefitting. Among others, some had interest in attending the programme, but due to personal and farm work, were unable to participate.

However, participants of these programmes come up with doubts and issues to ASHA worker and if she is unable to clarify, they take doctors' help either through phone calls or personal visits; 19% of the respondents rate home remedy programme as excellent, as it helps them use locally-available ingredients in the preparation of medicines. Most women expect a booklet of these remedies to be provided with as memorising/recalling the process is generally difficult, while 17% have rated the programme as excellent and good. This shows their interest to learn and acquire knowledge about various plants and their usage.

3.3.3 *Challenges*

a. ***Exclusive diagnosis done at DMRC, Soukya Foundation***

The doctors at DMRC, Soukya Foundation are specialised in their diagnosis and treatment. They initially identify the root cause of a given disease based on symptoms and through laboratory tests and if necessary, through scans (only in special or critical cases, which is not followed in most local clinics around Jadigenahalli) before deciding on treatment. The first step of analysis helps solve half of the problem; in the second step, medicines are prescribed according to the nature of disease and requirements; in the third and final step, the cause of the disease is identified with patients suggested to follow a particular diet, for a complete cure.

A Majority of the patients opine that medicines given by DMRC Soukya Foundation do not lead to any side effects, unlike Allopathy. Although these medicines take time to produce positive results and react at a slow pace, the impact lasts for a longer time. Homeopathy medicines provided by DMRC, Soukya Foundation, are easy to consume besides reducing stress on the part of parents, as children enjoy chewing them without resistance.

The number of patients approaching DMRC, Soukya Foundation, for treatment is on the rise (See Fig. 3.6). The reasons for this increase include the proximity to

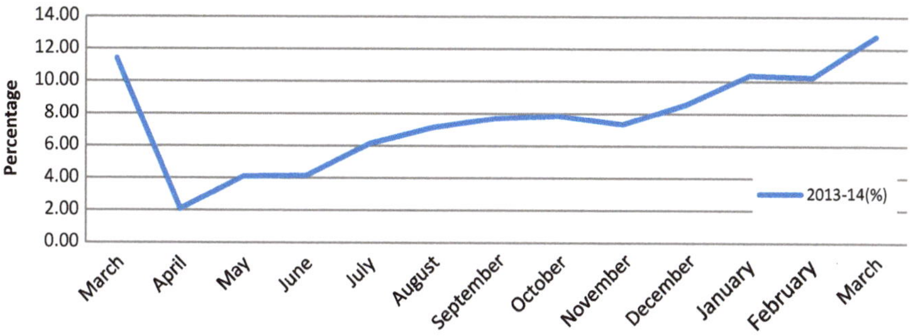

Fig. 3.6 Number of patients visiting AYUSH Grama DMRC, Soukya Foundation (2013–14)

Table 3.2 People's opinion about DMRC, Soukya Foundation according to ASHA worker

Programme	Jadigenahalli	Vadigenahalli	Govindapura	KB Hosahalli
Medicines	Good	Good	Better	Good
Treatment	Excellent	Excellent	Good	Good
Diagnosis	Good	Excellent	Very good	Good
Medical camp	Excellent	Good	Good	Excellent
Medicinal plants distribution	Good	Excellent	Good	Excellent
Usage of medicinal plants	Good	Good	Fair	Good
Home remedy	Excellent	Good	Good	Excellent

(Source: Authors 2014)

the Primary Health Care Centre, Ayurveda doctors working on a full time basis, new building constructed, all facilities under a single roof i.e. Yoga, Pharmacy, diagnostic lab, Ayurveda, Homeopathy and Naturopathy.

a. ***Stakeholders' Perceptions:***

(i) ASHA Workers

ASHA Workers play an important role and are one of the important catalysts of this programme who help convey the message to people at the local level. Hence, ASHA workers are identified and chosen with care, as it is important that they are aware of the objectives of the programme with certain qualities of communication and convincing skills. In this context, our discussions with the ASHA workers about the responses of people are explained below (Table 3.2). It is interesting to know that across the villages, the programme specifics are rated as excellent and good, excepting the usage of medicinal plants in Govindapura (rated as fair).

(ii) Village leaders

Locally, all village leaders are aware of DMRC, Soukya Foundation and its activities. The village leaders consider AYUSH as a good health care intervention benefitting people at large in terms of treating small ailments at low costs. The village elders are concerned with the gradual disappearance of traditional practices in the context of urbanisation affecting the lifestyles and belief systems. They welcome the revival of traditional systems of medicine for treating the ailments in place of allopathy.

The village leaders are also happy with the initiation of yoga as a regular activity as part of meeting health-related requirements of men and women. Men are free in the mornings and women in the evenings with their work patterns being suitable to performing yoga. The village leaders have also volunteered to provide a common place besides making required arrangements for conducting yoga classes.

(iii) Teachers

Teachers see this yoga programme as a good initiative in strengthening the already existing programme. There have been instances where children have participated in yoga competitions and won prizes. Teachers opine that it would be good if parents were involved in encouraging children, while at the same time, agree that parents do have practical constraints. For instance, in Govindapura, parents work as daily wage labourers and hence, will lose out on their daily earnings, if they attend camps or yoga programmes. There are other challenges such as fewer number of student enrolments, as parents prefer to send their children to private schools affecting the student strength in government schools.

3.3.3.1 Some Constraints

Extended discussions with DMRC, Soukya Foundation personnel enabled us to understand the following constraints.

- Administrative
- On the administrative front, getting doctors and post-graduates in social work to do a continuous service at DMRC, Soukya Foundation with a service attitude is challenging. Getting an uninterrupted flow of funds can be a challenge in the long-run. Hence, a workable model for sustaining cost effectiveness needs to be worked out. DMRC, Soukya Foundation finds it difficult to get the funding on time from the government. If money is provided in advance to ensure an uninterrupted flow of funds, work will not be affected. Major constraints faced include logistics and delays in the clearance of submitted bills. During the initial stages, there were delays due to the tedious and time consuming process, but it has been streamlined since September 2013. However, emergency cases cannot be handled as they lack facilities.

- People's Cooperation
- On the people's front, the challenge of influencing lifestyle changes is a tough task. In the initial stages, it took more time to create awareness as the AYUSH systems of medicine were new to the villagers. Diabetic conditions among several patients were identified for first. However, doctors faced difficulties in ensuring

lifestyle and dietary changes. This needs a continuous support from doctors through counseling and training. In the absence of proper treatment equipments, while treating those with high glucose levels are referred to better-equipped centres, but patients may not follow the advice. Hygiene and changes in habits take time; sometimes it becomes difficult to convince them. Approaching people to enroll for de-addiction programmes can be challenging as some people create disturbances, needing tact and patience to handle the situation. Many a time, if people are found unable to explain clearly the symptoms they are going through, it becomes difficult to treat. This could be due to poor educational levels. Initially, it was difficult to gain the confidence and attention of people in terms of getting them to participate in the programmes in spite of extensive publicity campaigns. However, Gram Panchayat members' cooperation is inadequate. Participation of working people is also found inadequate due to unsuitable timings. Currently, housewives are observed participating (30%) in various programmes being conducted, particularly those related to Home remedies.

3.4 Towards a Sustainable Approach

The current model of health care followed by DMRC, Soukya Foundation, needs to be sustainable in the long-run and replicable socially, economically and environmentally.

a. ***Financially Sustainable Options***
Patients could be charged depending on the economic group they belong to – below poverty line – BPL (free medicines), middle income group (moderate fees) and high income group (standard amount). Donations from other institutions could also be of help. The government could think of continuing the support for a fair length of time before the programme can sustain on its own. They could also provide free homeopathic medicines to a specified number of patients (like 100 patients) and thereafter at a subsidized price.

b. ***Strengthening Institutions and Integration***
The DMRC, Soukya Foundation, is a public-private venture set up to provide health services to the rural people. The Government of Karnataka, under the Ayush gram central scheme, provides Rs.34.95 lakh annually and has provided 3 acres of land for construction of the hospital and conducting the programme. There is a scope for involving other formal and informal institutions at local and regional levels that can strengthen the programme by complementing each other. For examples, involving schools and school children in various programmes, creating awareness by explaining the medicinal value of locally-available plants, growing medicinal plants, Yoga promotion through practice and initiation of youth, students and village leaders, on a monthly/quarterly basis and also programmes for promoting sanitation.

c. ***Networking for Outreach***
Currently, the programme is solely funded by the Government, with a fair scope for NGOs, corporate houses (through corporate social responsibility (CSR), religious institutions, educational institutions, philanthropists and research institutes to contribute to the programme. There could also be streamlined programmes focused on creating awareness regarding sanitation and hygiene or educational programmes for students at various levels and other targeted campaigns. For e.g. Titan (a Tata company) is interested in improving the lives of girl children; a financial organisation is interested in micro finance, while an NGO is keen on providing potable water.

d. ***Self-help groups***[11]
The programme can utilise the services of active women self-help groups under Jadigenahalli gram panchayat gradually at various levels. Currently, Jadigenahalli gram panchayat has 46 women self-help groups with some of them being already involved in business activities. For instance, Kaveriamma Swasahaya Sangha (an SHG) at Govindpura has been involved in social and environmental activities like keeping the surroundings clean by collecting waste materials, planting new seedlings, watering plants etc. Similarly, in Harlur, SHGs are into baby corn packing, while, in Jadigenahalli, a few SHGs are involved in vermin-compost preparation, indicating entrepreneurship at the group level.

All members of SHGs are aware of DMRC, Soukya Foundation, and its activities in their villages. They see it as a positive and useful initiative, as access to medical care and treatment has become easy and convenient. It is also free of cost, which is seen as a boon. They discuss these activities with their friends, neighbors and relatives, with a few of them attending medical camps, medicinal plant distribution drive and other activities. General perceptions and opinion about DMRC, Soukya Foundation, are quite positive across all SHGs, who consider the programme useful in terms of bringing about significant changes in the villages. Some of them would like to work as volunteers, if required, for DMRC, Soukya Foundation. However, they need a plan and specific tasks so that they can invest their time in its activities and also they require training.

3.4.1 Youth Organisations[12]

Youth organisations play a significant role at the village level. However, under Jadigenahalli gram panchayat, the number of youth organisations has declined due to urbanisation impacts such as well-educated youth migrating to cities. Currently, there are two youth organisations, willing to work with DMRC, Soukya Foundation, as social workers for spreading awareness regarding its activities,

[11] A self-help group (SHG) is a village based financial intermediary usually composed of 10–20 local women or men. Members make small regular savings contributions over a few months until there is enough capital with the group to begin lending. Self -help groups may be registered or unregistered as micro financial institutions for community action.

[12] Youth organisations in villages are private nonprofit organisations dedicated to helping emotionally and behaviorally-troubled villagers and their families.

particularly yoga and medicinal plant distribution. Currently, the organisations are involved in educational activities (offering rewards and recognition and also lending financial support to needy students), health camps (organising blood donation camps), and sports, camps besides their own yoga centres. Initially, in Govindapura, the focus was on curbing anti-social activities like illicit liquor sale, gambling, playing cards etc. Currently, they are focused on opening a gymnasium and a yoga center as well. However, Govindapura youth organisations prefer to spend a limited time on spreading awareness regarding medicinal plant distribution or other activities. The youth organisations meet regularly every month to discuss the status of various initiatives and to plan future activities that indicates the strength of institutions that can be tapped for an improved outreach. Both the youth organisations have agreed to work with DMRC, Soukya Foundation. JP charitable trust is interested in spreading awareness and information on yoga as they strongly believe that yoga is an effective option everyone can greatly benefit from. The Trust is also keen on converting their village into a green zone and hence, interested in planting more trees.

e. **Meeting Specific Preferences**

- ***Need for Lady Doctors***

  Currently, a large number of women visit DMRC, Soukya Foundation and hence, it's preferable to have a lady doctor since women generally suffer more from health-related issues, a majority of them feel more comfortable sharing their gynaecological and other personal concerns with women doctors.

- ***Need for Ayurveda Doctors and other Specialists***

  Generally, people are more aware of ayurveda than homeopathy and hence, faith in ayurveda is high. The villagers are generally inclined towards ayurveda given the age-old presence of traditional practices. A majority of them have requested for regular ayurvedic doctors, like homeopathy doctors. Also there are demands for gynaecologists and opthomoligists at least 3 days a week.

- ***Need for more Medical equipments for Diagnosis***

  Installing of more medical equipments like scanning machine, x-ray, blood check up instruments etc. and charging a discounted rate could make it convenient for patients. People are willing to pay for such services, though at reduced rates.

- ***Increasing the staff strength***

  The popularity of DMRC, Soukya Foundation, has led to an increase in the number of patients visiting it. People with major health ailments, who are unable to find a proper treatment elsewhere, visit DMRC, Soukya Foundation, as also people with minor health problems. Besides, patients with chronic problems like diabetes, hypertension and allergy-related ailments also visit the Foundation regularly. Positive words spread about the Foundation through relatives and friends of patients belonging to other villages/towns and even cities like Bangalore, resulting in an increase in the number of patients visiting the Soukya Foundation. Hence, it would be important to increase the staff strength to cater to the increasing demand.

- ***Advance Stock Procurement***

  Owing to increased number of patients and medicines, there is a focus required on an adequate stock. People observe that some times, the non-availability medicines has led to repeated visits, thereby to increase transaction costs, specifically for economically weaker sections.

- ***Strengthening Personnel's communication skills***

  The major role of ASHA workers is to motivate and increase awareness among people regarding AYUSH and also to help them get treatment at DMRC, Soukya Foundation, by strengthening the relationship between DMRC, Soukya Foundation and people. ASHA workers have expressed the need for further training in persuasion skills, communication skills, handling of varied situations etc. Besides these soft skills, some basic knowledge about diseases/ growing herbal plants/home remedies would also be useful. ASHA workers feel discouraged when some of the village folk hesitate/avoid/defer speaking to them. Also, getting people to participate in camps is a herculean task for ASHA workers. Similarly, home remedy programme is more in demand with a higher participation rate as compared to earlier days. People attending these programmes would like to have several doubts clarified on the spot, which necessitates strengthening ASHA workers' skills.

3.4.2 Improving Communication Strategies – Home Remedies in Printed Form

Home remedies taught in programmes should be distributed either in booklet form or printouts. Since it is difficult for old people to memorise all the items discussed, booklets in local language can be of good help.

3.4.3 Influencing People on Life Style Changes and Holistic Treatment

People in general are more used to allopathy medicines and because of an easy access to medicines at medical shops and also local shops (pain killers), self-medication is common. As pain killers provide an immediate relief, the harmful effects are not known to people. Besides, lifestyle changes mean that they have to make efforts by giving up various addictions and practising yoga, which is not convenient, given their mindset and also the day-to-day challenges of travel and work. Hence, time constraint is seen as an easy excuse. Besides, urban lifestyles largely influence them to approach private hospitals. In view of a close proximity to Bangalore city and diverse occupations, people generally enjoy a better economic status which also makes them approach private hospitals during times of sickness.

f. ***Ecological Protection and Conservation***

- ***Water Protection and Conservation Measures***

  Jadigenahalli gram panchayat falls under a semi arid zone, with a severe water scarcity. It is important to work on short-term and long-term solutions, while addressing this issue as part of ensuring the sustainability of all other related programmes. Rainwater harvesting to revive the existing tanks and smaller structures needs to be promoted. A proposal on sewage water treatment in Bangalore city is another feasible option. These are important from the view point of protecting the environment and health concerns of people. It is only by securing water availability that initiatives towards cultivating medicinal plants on a large scale could be taken up. Against the backdrop of a severe water scarcity, 86% of the farming families surveyed have expressed that it is difficult to continue with agriculture, given the water, space and also marketing constraints. Poor rainfall and increasing ground water extraction have led to serious concerns with respect to finances, livelihoods and agricultural crops.

3.4.4 Need to Promote Organic Farming

Farmers tend to use fertilisers and pesticides without bothering much about the undesirable consequences. Our observations show that farmers use pesticides without using any protection. They do not use masks and gloves while spraying pesticides on plants. This could cause serious health related ailments. Our discussions with farmers reveal that they are fairly aware of the negative implications of pesticide use, but tend to be negligent in protecting themselves. Gram Panchayat members and the elderly in the village express their concerns regarding the changing scenario of agricultural system, uncontrolled usage of fertilizers, depleting ground water levels and their implications for health and environment. Hence, they are keen on getting trained in organic farming, increasing awareness about harmful effects of fertilizer and pesticide use on health and environment.

3.4.5 Need for Improved Sanitation

Sanitation is another major concern in peri-urban areas. Solid waste management is a major threat faced by urban areas; the situation is changing in the peri-urban areas also due to changes in lifestyles and food habits. These changes normally lead to increased waste generation and related implications, especially for health. The generation of waste has increased; more so plastic usage. Our observations at the Jadigenahalli

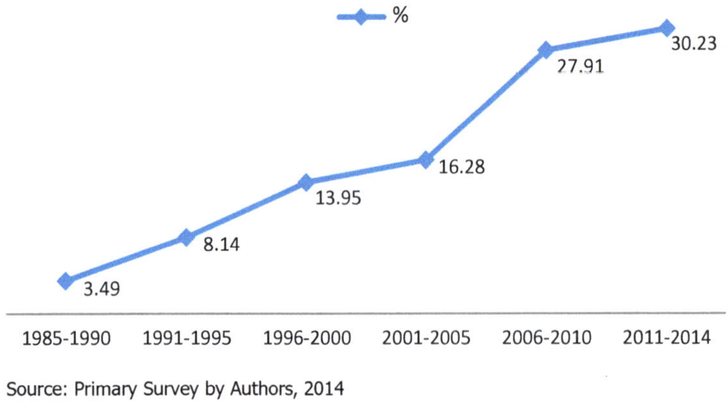

Fig. 3.7 Progress in toilet construction

gram panchayat indicate that there is no daily solid waste collection system in place, though it has made a provision for the collection of waste once in 6 months from designated spots where garbage is dumped by households, while 48% of the people are unsatisfied with the present system and 52% of the respondents are not in favour of the daily collection system by the Grama Panchayat, as they have their own spaces for dumping waste (organic waste), which they convert into farm yard manure (FYM) to be used in their own agricultural lands. In the usual course, villagers dispose off their organic waste and other litter waste each day. Some treatment methods commonly followed in these villages include open burning, dumping and selling. However, open burning releases smoke and other toxic elements into the atmosphere leading to air pollution, which the residents are not generally aware of.

Waste is loaded onto tractors to be transported to the vacant government lands, specified quarries and dumpsites. Since Jadigenahalli is closer to the peri-urban town Hoskote, urbanization and development largely influence the village population towards the construction of houses for commercial purpose. This has resulted in a reduction in vacant sites for disposal of waste. The most common problem facing urban or peri commercial establishments is solid waste disposal due to the lack of a sufficient space. In order to solve this problem, the commercial establishments have started dumping their waste in and around the villages under Jadigenahalli Panchayat. These are common scenes witnessed in a majority of the peri-urban areas around Bengaluru.

Another major concern is the underground drainage. While some of the peri-urban areas have access to an underground drainage system, a majority go without. In 2011, the Jadigenahalli Grama panchayat received Suvarna Grama Yojane Award from the Central Government for their excellent work related to sanitation, cleanliness, hygiene and usage (Fig. 3.7). These awards are given to villages with a population of more than 2000 and people's access to toilets. Prior to toilet construction, constraints encountered were pathetic.

A significant influence of urbanization can be seen in the construction of private toilets, as urban residents are more aware of cleanliness and prefer to have own toilets. Children are educated in urban areas and are keen on having private toilets. That literacy percent stands at 71.02%, can be seen as one of the major factors influencing mindset of people towards the construction and usage of toilets (61.76% usage and 22.55% non usage). Here non usage category is men and senior citizens, because they are not used to closed sanitation. They still prefer open defecation, while 57% of the respondents have expressed that the construction of toilets protects their privacy and dignity. Women are convinced of increased safety (44%) and comfort (43%). Women feel the need for toilets the most, and are, therefore, the biggest beneficiaries when they are constructed, while 32% have reported improved hygiene in the village and 17% an improvement in households have reported with a sufficient space and economically better off have constructed their own toilets.

Awareness programmes regarding health and hygiene need to be promoted. Currently, toilet construction is nearly complete and hence, there is only a small percentage of people practicing open defecation. This could be curtailed completely. Besides this issue, a greater concern relates to solid waste management. Garbage is strewn all over the village besides designated spots (largely open sites and vacant places) resulting in poor aesthetics and hygiene in the village. People burn garbage, which is more harmful in terms of causing air pollution. There is a need for streamlining garbage management. The initiative taken upby the gram panchayat has to be further strengthened for managing garbage.

3.5 Summary

Healthcare is the right of every individual, but lack of quality infrastructure, dearth of qualified medical functionaries, and non-access to basic medicines and medical facilities limits its reach to 60% of the population in India. Considering the challenge of providing health care in India, with rising costs and infrastructural issues, traditional medicine is seen as a positive option for improving the health of masses. Thus, DMRC Soukya Foundation as an intervention of AYUSH programme by the government has had a positive impact on economically-poor communities. However, public-private partnerships cannot survive in the long-run without financial sustainability. To become self-financing there are two possibilities: making the patients pay. Possible financial options include for varied economic groups could be – BPL (free medicines), MIG (small fee) and HIG (standard amount). Building a corpus of funds from donors like corporate houses under Corporate Social Responsibility (CSR) programs, religious institutions and philanthropists. The Soukya-DMRC initiative is a shining example of how luxury spas offering services based on traditional medicines can also run hospitals for the rural poor as part of their corporate social responsibility. In sum, provision of quality traditional medicines-based healthcare seems a very good option for improving access to health care in rural India and other developing countries. Public-private partnerships can be an effective vehicle for promoting the same. However, their financial sustainability after the period of partnership tenure has to be reflected upon, and even decided, before implementation.

Annexures

Annex 3.1 NGOs involved in Ayush Grama

Sl.No	Name of NGO	Project area	District
1	Gramothan foundation, Bangalore,/Surya Foundation Mysore	D B Kuppe	Mysore
2	Vivekananda Girjanara, Kalyana Kendra	Yargumnahalli	Chamarajnagara
3	JSS Mahavidya Peetha	K Shetty Halli	Mandya, Mysore
4	Soukya foundation	Jadigenahalli	Hoskote, Bangalore rural
5	Sri Sri Ayurveda Viganamahavidyalaya, Bangalore	Kalahalli	Ramanagara
6	Vivekananda yoga Anusandana Samsthe, Bangalore	Khanagoan	Belagaum
7	Nisarga trust, Sirsi	Baiumbe	Uttara Kannada
8	Shanthivana trust, Dhramastala	38th Kalthur	Udupi
9	Parivarthana Parivara, Belgaum	Peeanwadi	Belgaum
10	Bellur Krishamachari Sessamma Smaraka Nidhi trust, Bangalore	Bellur	Kolar

Source: http://www.sscasrh.org/index.php/achievements/item/231-ayush-grama-another-feather-in-the-cap-6th-feb-2013

References

C. Wang, Holistic Health Definition: The Essential Elements, Miami FL, April 27th 2017 in https://www.amcollege.edu/blog/the-essential-elements-that-define-holistic-health, (2017)

A. Costello et al., Managing the health effects of climate change, Lancet and University College, London Institute for Global Health Commission. The Lancet **373**, 1693–1733 (2009)

P.P. Mhame, K. Busia, O.M.J. Kasilo Clinical practices of African traditional medicine, The African health monitor, Special issue 14, Decade of African Traditional Medicine, 2001–2010, (2010).

P.P. Mhame, K. Busia, O.M.J. Kasilo (2001 to 2010), 'Clinical Practices of African Traditional Medicine, (2001–2010), 1 Traditional and Alternative Health Practice Council, Ministry of Health and Social Welfare, Dar es Salaam, United Republic of Tanzania 2 West African Health Organization 3 WHO Regional Office, Decade of African Traditional medicine, special issue in http://www.aho.afro.who.int/en/ahm/issue/13/reports/clinical-practices-african-traditional-medicine, African Health Observatory

R.K. Pachauri, A. Reisinger, and IPCC (2007). Climate Change 2007: Synthesis report, contribution of Working groups I, II and III to the fourth assessment report of the intergovernmental panel on climate change, Geneva IPCC

A. Pruss – Ustun, J. Wolf, C. Corvalan, T. Neville, R. Bos, M. Neira, Diseases due to unhealthy environments: an updated estimate of the global burden of disease attributable to environmental determinants of health. J. Public Health **39**(3)., September), 464–475 (2017)

D. Shankar, B.K. Patwardhan, B. AYUSH for new India: Vision and strategy. J. Ayurveda Integr. Med. **9**(3), 137–139 (2017)

U. Payyappallimana, Broadening perspectives of global health – prospects for AYUSH and integrative medicine. J. Ayurveda Integr. Med. **9**(2), 87–89 (2018)

WHO, World Health Assembly Discussions held annually by the WHO, (2018)

Websites

http://www.channel4learning.com/sites/gcsease/health_social/2_1.html
http://www.alternativehealth.ie/a-holistic-approach-to-health-and-well-being
http://www.nationalaccountsofwellbeing.org/learn/need/
https://www.ancient-origins.net/ancient-places-americas/what-ancient-mayans-can-teach-us-about-health-and-healing-001220 - Accessed on 23-1-19
http://www.shen-nong.com/eng/principles/holism.html
https://www.bing.com/search?q=ayushman+bhava+bharat+scheme&form
https://ahha.org/selfhelp-articles/holistic-health/

Chapter 4
Cultural Influences on Health, Traditions and Ecology

Abstract The chapter looks into rural ecology and its implications for people's lives and environmental conservation. It examines the inter-linkages between ecology and cultural traditions followed by people in their day-to-day life with implications for climate change. It also explains how traditional knowledge has all along played a vital role in maintaining the health of people through rituals and festivals, using medicinal plants. In order to understand the above, information on the natural resources, cultural traditions and ancestral knowledge was collected through focused group discussions, and informal talks mainly with knowledgeable local elders and women. Based on the study, it was clear that the cultural practice of worshiping nature has enriched their health, local culture and traditions, which have helped conserve natural resources and medicinal plants, in turn, captures the dimension that culture plays a key role in reducing climate change impacts.

Keywords Culture · Traditions · Natural resources · Climate change · Health

4.1 Background

Cultural dimension has an immense on climate change impacts and adaptation. Cultural values inspire people to express concerns over a range of social and environmental problems, and such morals are linked to their actions to address these problems. Jessica Barnes et al. (2013) argue that understanding the challenge that climate change poses and crafting appropriate adaptation and mitigation mechanisms requires inputs from both natural and social sciences. Anthropology's in-depth fieldwork methodology, long engagement in questions of society–environment interactions and a broad, holistic view of society capture valuable insights into the science, impacts and policy of climate change. Similarly, Andrew Mathews and Jessica O′ Reilly (2013), also suggest the collaboration of experts from both natural and social sciences for addressing the mammoth challenges posed by climate change. Further, the authors identify three ways that anthropological research can aid, enrich and deepen contemporary understandings of climate change. To begin with, the discipline's tradition of in-depth fieldwork gives anthropologists insights

S. Manasi, K. V. Raju, *Coping Mechanisms for Climate Change in Peri-Urban Areas*, https://doi.org/10.1007/978-3-030-18517-6_4

into the cultural values and political relations that shape the production of climate-related knowledge. Secondly, it is an awareness of the historical context supporting contemporary climate debates. Thirdly, it is anthropology's 'broad, holistic view of human and natural systems, which highlights the multiple cultural, social, political and economic changes that take place in our societies'.

Besides, American Anthropological Association (2014) has come up with a report by the AAA Task Force on Global Climate Change, along with a Statement on Humanity and Climate change. Shirley J Fiske (2015), Chair of the American Anthropological Association Global Climate change Task Force, states that 'Anthropologists focus on several aspects of climate change research that other scientists do not fully address, specifically the disproportionately adverse impacts on vulnerable populations, the extent to which our current challenges stem from culture and cultural choices on a societal level; and the value of the long record of human development and civilization that can inform our choices for the future'. The statement of the report frames the global challenges of climate change as being rooted in social institutions and cultural habits, demonstrating that solutions and social adaptations consequently require knowledge and insights from social sciences and humanities (for the same reason, this study related to JDGP, looks at the social adoptions at the local-level and link them to cultural habits, as described in the following sections). Ben Orlove (2015), task force member, stated that 'Resilience and adaptation can be best addressed locally and regionally, by enabling communities to provide knowledge and social capital to construct viable solutions'. The report wraps up highlighting that top-down policies and programs may not be the paramount way of enabling communities to address the effects of climate change. While climate change will have a global impact, the impact will fall unevenly, and as climate impacts intensify, public expenditures needed for emergency aid and restoration will also shoot up.

While Andrew J Hoffman (2015), in his book states that, though the scientific community largely agrees that climate change is underway, debates about this issue remains fiercely polarized. He then extracts lessons from major cultural shifts in the past to engender a better understanding of the problem and to motivate the public to take action. 'How culture shapes the climate change debate' lays out the psychological and social biases people bring to the climate discussion and then suggests techniques for making that dialogue more productive. He asserts that the public perception of climate change debate is rooted in avoidance, thus, focusing on values not science. Besides, Hoffman (2015) also describes the climate change debate at its core as 'cultural' and that social scientists view the public understanding of climate change not as a lack of adequate information but as an intentional or unintentional evasion of that information. That evasion is rooted in our culture and psychology. Social scientists find that physical scientists do not have the final word in public debate. Information is searched and reached to conclusions about highly multifaceted and politically contested issues in a way that will direct them to find supportive facts from one's pre-existing beliefs. Cognitive filters reflect cultural identity. Cultural identity can overpower scientific reasoning.

The debate over climate change is not about carbon dioxide and green house gas models, it is about opposing cultural values and world views through which science is seen. The social debate around climate change is now devolving into a 'cultural schism', in which opposing sides do not debate the same issues, seek only information that supports their position or disconfirms the other's and begin to demonize those who disagree with them. With time, our positions become relatively rigid and exclusive, thickening the boundaries between cultural communities.

It is now increasingly being recognized that nature and its rich biodiversity still support, and in turn, are maintained by a great diversity of ecosystem people and their cultures the world over (Gadgil 1995; McNeely 1995). The hunter-gatherers and shifting cultivators throughout the world typically possessed a 'weltanschauung' in which man is considered an integral part of the ecosystem. Such a worldview does not profess the concept of man-nature dualism, as emphasized in the western philosophical thoughts of Descartes, Bacon, Liebnitz, and many others. The 'organic cosmology' of the indigenous societies shaped an ecological ethic that is undermined by western science and market-oriented culture (Merchant 1980; Nelson 1993). Nevertheless, those indigenous cultures that are still surviving retain their traditional ecological ethic, which now seems to have profound conservation implications (Gadgil and Guha 1992).

The Northeastern region of India is home to a myriad of tribal and other ethnic groups. Large sections of these communities can be still said to be earning their livelihood by exploiting their biomass resources as 'ecosystem people' (Dasmann 1988; Gadgil 1995). Landauer et al. (2014) deals with the specific concerns and highlights the significance of culture; investigates cultural differences in tourist preferences for climate change adaptation strategies of cross country skiers in Austria and Finland. They use the value orientation approach to testing empirically whether this concept is sensitive to skiers' climate change adaptation preferences in the two respective countries. The comparisons between the two countries are made even more specific with three identical motivation-based segments that account for heterogeneity within the samples. The results reveal significant differences between skiers for the overall national samples, as well as for numerous comparisons between specific segments. Thus, it concludes that adaptation strategy planning should consider cultural differences and that the transferability of strategies even within the western cultures of Europe is limited.

Thus, it is viewed that we need a more rigorous approach to tackling social and environmental challenges, the path which understands the morals/ethics and the interconnectedness of these morals and behavior and the ways in which some of these are reinforced culturally. The response to climate change also depends on the level of aspiration required for effectively addressing the challenge. There are indicative evidences from social psychology that highlight the importance of cultural values in shaping collective responses to challenges.

Local field work facilitates such perspectives to be critically assessed (Roncoli 2006). Several studies conducted locally have pointed to indigenous methods of predicting several aspects like for instance, rainfall, rain guage data showing an accurate institutional memory of rainfall events. Understanding the 'human dimensions' of inter-annual climate variability' and local responses is a vital theme in 'climate anthropology' (Vasquez Leon et al. 2003; Finan and Nelson 2001; Magistro and Roncoli 2001; Roncoli et al. 2001; Roncoli 2006; Strauss and Orlove 2003). Climate knowledge is conveyed inter-generationally and across social networks, but new weather patterns render it indistinct. Scientific forecasting can now assist farmers and herders in remote regions by providing event probabilities, but understanding how this is interpreted and understood is an ethnographic task (Brown 1999). Climate anthropology operates in multiple knowledge communities and bridging these remain vital.

Given the detailed discussions and broader dimensions that indicate the role of culture in addressing climate change, in this chapter, we have identified the linkages between cultural traditions in practice and their influence in revitalizing health and protecting the local environment and in larger contexts – climate change. The objective was to understand the role of cultural traditions/practices in the conservation and usage of medicinal plants in relation to protecting the health status of people. We captured practices related to religious beliefs, cultivation of medicinal plants and their usage, ecology and cultural aspects in-depth. We have found the adaptation to a holistic health care of DMRC Soukya Foundation's intervention, discussed in our previous chapter, being rooted had its influence in the cultural traditions in revitalizing health and also conserving the local environments. Across the world, cultural practices among communities are diverse and nature worship is popularly practiced. Nature is being worshipped to ensure the welfare and health of human beings. There are strong evidences indicating linkages between environment and health since thousands of years, as reflected in the fourth century BCE Greek scriptures, Indian Upanishads, Chinese and Western Asian and European literature (Guha 2002). Traditional knowledge is important for its own sake as also for its social and cultural values, but it is also significant for a number of practical reasons relevant to biological diversity (IUCN report 1986). Indian nationalism has grown out of an amalgamation of scores of nature-based local cultures and practices, evident as it has been from the worship of plants, groves, animals and natural resources like rivers, mountains, ant-hills and rocks (Ramachandra 2008). Much of the world's biodiversity has been in the hands of traditional people, societies of hunters and gatherers, herders, fishers, agriculturists for a great many generations (Gadgil et al. 1995).

In a study on tradition and conservation in northeastern India, Abhik Gupta and Kamalesh Guha (2002) observe that nearly 90% of the conservation of wild animals and birds in Assam, Mizoram, Tripura, Manipur, Megalaya is due to traditions and cultural practices followed by local and various tribal communities. In a study on Sacred Groves by Oliver King et al. (1997) observe that at the entrance of sacred groves in Cacher, and Assam, people are required to remove their footwear. This means showing respect to nature and protecting the sensitive species of the region as well. The fact is, the pre-historic and scientific, traditional systems of management have been the main means by which societies have managed natural resources for millennia (Berkes and Farvar 1989; Gadgil et al. 1993). In many cases, the main reason we have any biological diversity to speak about is these systems of management (Gadgil et al. 1993).

In many parts of India, communities have inherited the rich tradition of love and reverence towards nature through ages. Religious preaching's, traditions and customs have played a big role in this regard. Indian religious scriptures have generally been the advocates of environmentalism (Ancient Indian scriptures like Atharva Veda,[1] Mahabharatha,[2] Lalitavistara Manuscript,[3] Bhagvatha puranas[4] speak of the protection of earth through worshipping it as Mother). The epic Mahabharata has references made to the worshipping of cosmic Tree (pipal tree) in that those who worship it are regarded as worshipping the whole universe with gods, demons, animals and human beings. With respect to the Indian traditional knowledge on environmental conservation, a survey conducted by the Ministry of Secondary Education in India points out that through poems, songs and moral stories, the traditional knowledge or ancestral knowledge regarding worshipping and the using of medicinal plants was transferred to the next generations. The study also highlights that sculptures, scripts and monuments that we see today played a vital role in transforming the traditional ecological knowledge besides building faith and respect among the young minds with regard to historic times (Naga-Nagini Sculpture in Belur, Faith and ecology). Kalpavruksh Environment Action Group, Pune, highlights the protection of trees in ancient times through artistic forms like sculpture. Environmental governance as described in Kautilya's Arthshastra[5] highlights the need for conserving forest sanctuaries and punishing those who violate the rules by hunting animals or cutting down trees and plants (Indian traditional knowledge on Environmental conservation-Module, Ministry of Education, India).

Thus, traditional practices have contributed towards the conservation of medicinal plants and protection of health of the local communities. These practices include domestication, sacred beliefs regarding tree/plant worship, and respect towards cultural forests, cultural ponds and so on. However, there has been a decline observed in the number of medicinal plants over time due to several factors such as an increasing demand for primary health care, trade, food, timber and firewood. Besides, land clearance for agriculture, human settlements, increasing urbanization, forest fires also have contributed significantly to the loss of diverse species. Increasing urbanization is one of the major reasons, which has impacted changes in the cultural practices leading to a decline in culture and traditions, as also traditional health practices. Changes in life style, environmental pollution and degradation of natural resources also have close linkages to the deteriorating health of the masses. The current study[6] shows that the role of traditional management practices is pertinent and important in

[1] One of the four Vedas- *Atharvan*, (an ancient Rishi, and veda, meaning "knowledge") is a sacred text of Hinduism.

[2] one of the two major Sanskrit epics of ancient India.

[3] "The Play in Full" or "Extensive Play," referring to the Mahayana view that Buddha's last incarnation was a "display" or "performance" given for the benefit of beings in this world.

[4] Literally Meaning *Divine-Eternal tales of Supreme God-* great texts of Hinduism with their focus on religious devotion to Supreme God, primarily focusing on Krishna.

[5] The Arthashastra is an ancient Indian treatise on statecraft, economic policy and military strategy, written in Sanskrit.

[6] The current study refers to cultural traditions in conservation and Usage of medicinal plants in enabling better health in peri-urban areas.

protecting the rural ecology and health of the communities through cultural practices of biodiversity conservation and related health practices. In this context, it would be important to focus on self-management systems in respect of the future in regions where systems are diverse and scientific data are limited, through encouraging cultural knowledge in local ecological knowledge.

4.2 Local Ecology and People's Health

Traditional practices of protecting and treating health-related issues, using the locally available resources (plants and ingredients used at home) continue to be prevalent even to this day, passed on through generations mostly by word of mouth and practice. However, one can observe a lot of changes in these practices due to influential factors in the current contexts. An easy accessibility to allopathic medicines and off-the-counter purchase of medicines are also responsible, to a large extent, for these changes. In some cases, despite having medicinal plants in their backyards, they do not care to use them; rather they prefer to purchase allopathic medicines. Besides, although many of them are aware of the varied uses of medicinal plants and their healing properties, they are not able to grow them due to space and water constraints. But it is interesting to observe that a majority of the households in the study villages have several medicinal plants grown in their backyard gardens (Table 4.1), which play an important role in the conservation of environment and maintenance of human health.

Table 4.1 Medicinal plant usage and practices

SN	Usage and practices	Significance/Usage for ailments
1	Coriander seeds, turmeric, tulsi, ginger, milk and jaggery are boiled together to make a decoction for children	Cold, cough, fever
2	Doddapathre with salt is used for treating cough	Cough and respiratory problems
3	Tulsi with honey is used for treating cold and cough	Cold, cough and also strengthens immunity
4	Tulsi with salt and cloves is used for cold and cough both for children and adults.	Cold, cough, fever
5	Doddapathre, tulsi, kuppe leaf covered with castor leaves are burnt in a Chula, then taken out and squeezed into a liquid, mixed with milk or with lactating mother's and consumed; also used for severe cough	Cold, cough, fever in infants
6	Alovera paste is applied for healthy skin and healthy scalp.	Women and children common ailments, menstruation pains.

Source: Authors Field Survey 2014

4.2.1 *Common Ailments*

Seventeen diseases were found commonly affecting people in this Semi-arid tropical region (Fig. 4.1). As the main occupation of the people is agriculture and livestock rearing, it is labor-intensive and causes physical exertion and body aches; 23% of the people suffer from severe body pain followed by Diabetes, gastricitis, fever, skin allergy and Arthritis[7].

4.2.1.1 Medicinal Plants, Varieties and Usages

India has a rich culture of medicinal herbs and spices with more than 2000 species over a vast geographical area with high a potential for Ayurvedic, Unani, Siddha traditional medicines, though only a very few have been studied chemically and pharmacologically for their potential medicinal value (Gupta et al. 2005; Sandhu and Heinrich 2005). Human beings have used plants for the treatment of diverse ailments for thousands of years (Sofowara 1982; Hill 1989). According to the World Health Organization, most populations still rely on traditional medicines for their psychological and physical health requirements (Rabe and Van Stoden 2000), since they cannot

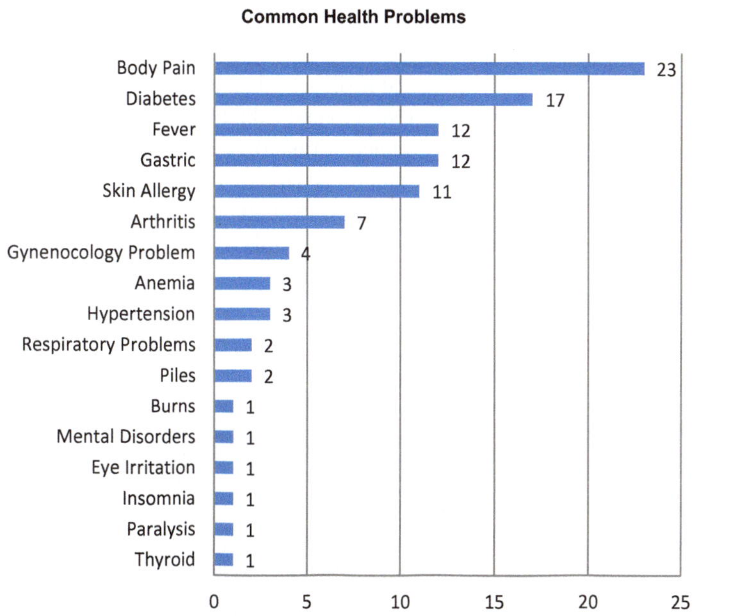

Fig. 4.1 Common health problems

[7]This is based on the patient's register data of Soukya Foundation Holistic Centre, Jadigenahalli.

afford the products of Western Pharmaceutical industries (Salie et al. 1996), together with their side effects and lack of healthcare facilities (Griggs et al. 2001). Rural areas of many developing countries continue to rely on traditional medicines for their primary health care needs. These medicines are relatively safer and cheaper than synthetic or modern medicines (Iwu et al. 1999; Idu et al. 2007; Mann et al. 2008; Ammara et al. 2009). People living in rural areas, based on their personal experiences, know that these traditional remedies are a valuable source of natural products for maintaining human health, though they may not understand the science behind these medicines, but know that some medicinal plants are highly effective only when used in therapeutic doses (Maheshwari et al. 1986; Van Wyk et al. 2000).

Medicinal plant cultivation has been prevalent across the region with households extensively using medicinal plants both for medicinal purpose and dietary and cultural practices (Fig. 4.2). Besides, the DMRC Soukya Foundation's intervention also has been promoting awareness regarding medicinal plant usage through their home remedies program and medicinal plant cultivation which, in turn, has further strengthened the usage of medicinal plants by villagers.

Interestingly, all the study villages cultivate medicinal plants (Table 4.2) with households in Vadigehalli accounting for the highest percentage in terms of the cultivation of medicinal plants (83%), while the least amount of medicinal plants being grown in Govindpura and Haraluru villages is mainly due to water scarcity.

Varieties of medicinal plants Among 14 varieties of medicinal plants, more commonly grown, Basil (Tulasi), Mint (Pudina), Indian borage (Doddapatre), Hibiscus (Dasavala), Tinospora Cordifolia (Amruthballi), Withania Somnifera (Ashwagandha), Aloevera, Curry leaves, Drumsticks (Nugge) and Adhatoda (Adusoge) were selected for a household survey. Calotropis gigantean (Yakaa), Neem (Bevu), Lemon (Nimbe), Mango (Maavu), Ginger (shunti), Cantella Asiatica (Vandelaga/Brahmi), Withania Coagulans (Panner), Epiphyllum Oxypetalum (Brahma Kamala), Garlic (Bellulli) are also grown, but comparatively on a small scale (Table 4.3).

Fig. 4.2 People growing medicinal plants

Table 4.2 Medicinal plants grown in Jadigenhalli gram panchayat

S n	Local name of plant	Plant species		Sample of plant	Sample of leaf	Parts used	Uses	Benefits
1	Aadusoge/ Adathoda vasica	Botanical name	*Adhatoda beddomei* C.B.CLARKE			Leaves	Respiratory problems and bronchial asthma	Health
		Family	ACANTHACEAE					
		Threat status	Critically Endangerd/ Global					
		Used in	Ayurveda					
2	Amruthaballi/ Guduchi	Botanical name	*Tinospora cordifolia* (WILLD.) HOOK.F. & THOMS.			Stem	Fever, urinary disorders, dyspepsia and urinary diseases	Health
		Family	MENISPERMACEAE					
		Used in	Ayurveda, Folk, Homeopathy, Tibetian, Unani and Sidha					
3	Aresena/ Turmeric plant	Botanical name	*Curcuma domestica* (MEDIK) VALH			Roots	Food, Folk	Health and spiritual and dietary
		Family	ZINGIBERACEAE					
		Used in	Folk					
4	Ashwagandhi/ Indian ginseng	Botanical name	*Withania somnifera* DUNAL			Roots	Nerves disorder	Health
		Family	SOLANACEAE					
		Used in	Ayurveda, Folk, Homeopathy, Tibetian, Unani and Sidha					

(continued)

Table 4.2 (continued)

S n	Local name of plant	Plant species		Sample of plant	Sample of leaf	Parts used	Uses	Benefits
5	Bettada nelli/ Amla	Botanical name	*Emblica officinalis* GAERTN.			Fruits	Ayurveda, Folk, Tibetian, Unani and Sidha	Medicine
		Family	EUPHORBIACEAE					
6	Dasavala/ Hibiscus	Botanical Name	*Hibiscus rosa-sinensis* L.			All parts	Hair conditioner,Uninorm (preventing irregular mensatio), sickness	Health
		Family	MALVACEAE					
		Used In	Ayurveda, Folk, Unani and Sidha					
7	Dhavana	Botanical name	*Artemisia pallens*.			Leaf with twig	Aromatic in nature	Ornamental plant
		Family	Asteraceae					
		Used in						
8	Dodabasle/ Spinach	Botanical name	*Basella alba* L.			Whole plant	Ayurveda and Sidha	Health
		Family	BASELLACEAE					
		Used in	Ayurveda and Sidha					

9	Doddapathri/ Indian borage	Botanical name	*Plectranthus amboinicus*			Whole plant	Cold, cough, fever and head ache	Health
		Family	LAMIACEAE					
		Used in	Ayurveda, Folk and Sidha					
10	Karibevu/ Curry leaf	Botanical name	*Murraya koenigii* (L.) SPR.			Root, bark, leaves	Sickness, nausea and bilious	Health
		Family	RUTACEAE					
		Used in	Ayurveda, Folk and Sidha					
11	Kaththale/ Cantala	Botanical name	*Agave cantala* ROXB.			Whole plant	Ayurveda, laxative, wound healing, skin burns, ulcer and cosmatics	Health
		Family	AGAVACEAE					
		Used in	Ayurveda					
12	Krishna Tulasi/Tulasi	Botanical name	*Ocimum basilicum* L.			Leaves	Antiviral agent, cough	Health and spiritual
		Family	LAMIACEAE					
		Used in	Ayurveda, Folk, Homeopathy, Tibetian, Unani and Sidha					
13	Nelli/Amla	Botanical Name	*Emblica officinalis Gaertn*			Fruits	As food	Health
		Family	Phyllanthaceae					
		Used In	Ayurveda, Folk and Sidha					

(continued)

Table 4.2 (continued)

S n	Local name of plant	Plant species		Sample of plant	Sample of leaf	Parts used	Uses	Benefits
14	Pannerelle	Botanical name				Leaves	As Ornamaental and cosmetic use	Health and spiritual
		Family						
		Used in	Aromatic and ornamental plant					
15	Pudina/Mint	Botanical name	*Mentha arvensis* L.			Leaves	Dietary food, gastric, indigestion	Health
		Family	LAMIACEAE					
		Used in	Ayurveda, Folk, Unani, Sidha and modern					
16	Sree Tulasi/ Tulasi	Botanical name	*Ocimum sanctum* L.			Leaves	Antiviral agent, cough	Health and spiritual
		Family	LAMIACEAE					
		Used in	Ayurveda, homeopathy, Tibetian, Unani, Sidha and Folk					

Source: Authors 2014

Table 4.3 Medicinal plants grown in percentage

Villages	Tulasi	Mint	Doddapatre/Thick leaved lavender	Hibiscus	Amruthballi	Ashwa gandha/ Indian ginseng	Aloevera	Curry leaves	Drum sticks
Govindpura	50%	28%	11%	28%	5.50%	0	22%	27.70%	5.50%
Haraluru	50%	17%	27.70%	33%	33%	0	33%	33%	5.50%
Jadegnahalli	69%	7%	54.70%	43%	28.50%	2.30%	47.60%	38%	9.50%
KB Hosahalli	66.60%	1%	25%	33%	16.60%	0	33%	0%	0.00%
Vadigehalli	50%	1%	66.60%	50%	66.60%	0	41.60%	50%	0.00%
Total	57%	11%	37%	37.40%	30%	0.00%	35%	29.70%	4.10%

Source: Authors 2014

The most commonly and largely grown medicinal plant across households is Tulasi (57%), followed by Doddaptre, aloevera and hibiscus (37%) and Amruthballi and Curry Leaves (30%), Mint Drumstick, Ashwagandha (Indian Ginseng, also known as Winter Cherry), Yakka, Neem, Lemon, Mango, Ginger, Vandelaga, Brahmi, Panner, Brahma Kamala, Garlic. These plants are used in their diet and for medicinal purposes. Turmeric plants are found only in Jadigenhalli and K B Hosahalli and are used for domestic purposes. Adusoge has been introduced by Soukya foundation. Amruthaballi is grown on a large scale, used by diabetic patients and hence, almost all houses with diabetic patients have this plant grown in their backyards, which they consume regularly as their dietary supplement (1–2 leafs per day directly or in liquid form). Curry leaves[8] and drum sticks are grown for both domestic and commercial purposes. The variations in medicinal plant varieties grown depend on peoples' interests, availability of plant species in the village, resources, Water, village-level practices, number of elderly people, influence of urbanization and so on.

Medicinal plant usage Since ancient times, in search of remedies for their diseases, people have been looking towards nature. The beginnings of medicinal plant use were instinctive. The Indian Holy Scriptures, Vedas, make a mention of treatment with plants, available abundantly in the country. Numerous spice plants used even today originated from India: nutmeg, pepper, clove, etc. Healing ailments with medicinal plants is as old as mankind itself. In the study villages also, it is seen, there is a high usage of medicinal plants as shown in Figs. 4.3 and 4.4.

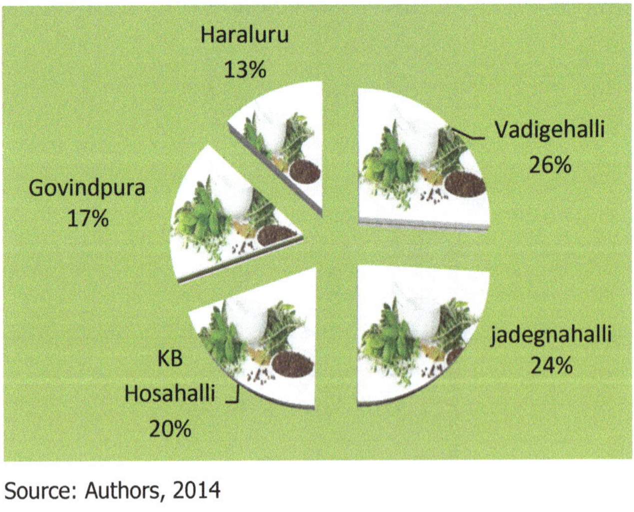

Fig. 4.3 Medicinal plant usage (%)

[8] In Jadgenhalli, it is seen that more number of curry leaves are imported and grown with a huge market for this plant. Curry leaves provide income for housewives (women empowerment) who purchase it in bulk and make into small bundles to be sold in the nearby market (K R Puram and Bangalore).

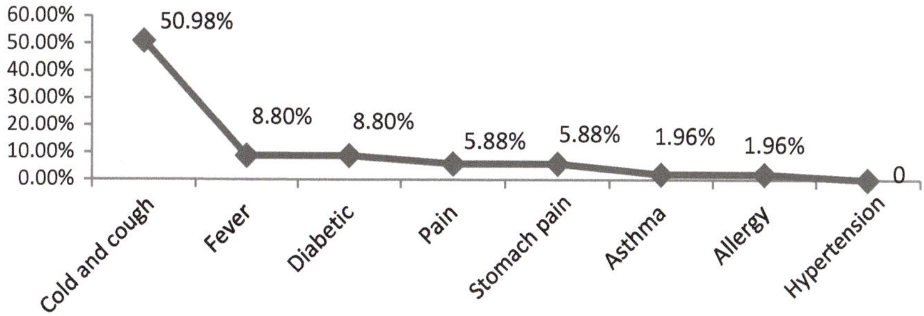

Source: Authors, 2014

Fig. 4.4 Usage of medicinal plants for different ailments (% share)

Fig. 4.5 Parts of medicinal plants used (%)

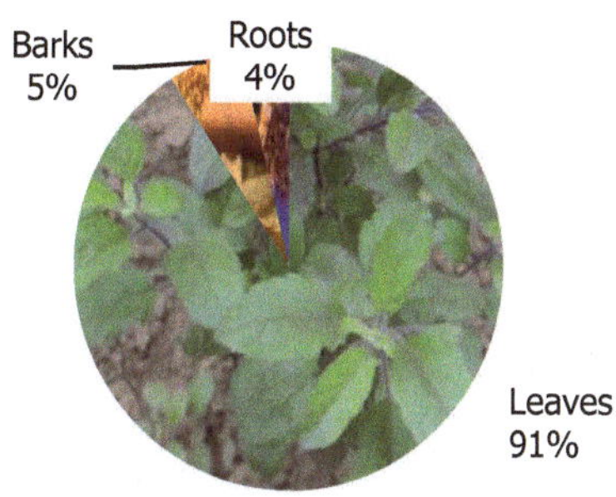

Source: Authors, 2014

People use medicinal plants depending on the severity of ailments. A majority of the people (51%) use medicinal plants for treatment of cold, cough and other ailments in varied proportions. Leaves of Tulasi, Doddapatre, Mint, Curry leaves are used (49%), while roots and Barks of plants used in making medicines account for 4% and 5%, respectively (Fig. 4.5).

Some medicinal plants are consumed directly, while some are processed to be used in the form of decoction, paste, powder, as specified for different ailments; 36% is consumed directly in the form of leaves, while 28% in the form of decoction. A few barks of medicinal plants including turmeric, garlic, ginger are used by grinding them into a paste (6%); Coconut, Sunflower, Ground nut, Nilgiri are used in oil form in dietary food and for medicinal purposes (7%). Neem paste along with onion and turmeric is also used for treating cut-wounds. Turmeric, coconut oil and horse gram are used for treatment of skin allergies (Fig. 4.6).

**Decoction* is a method of extraction by way of boiling dissolved chemicals from herbal or plant material, which may include stems, roots, barks and rhizomes

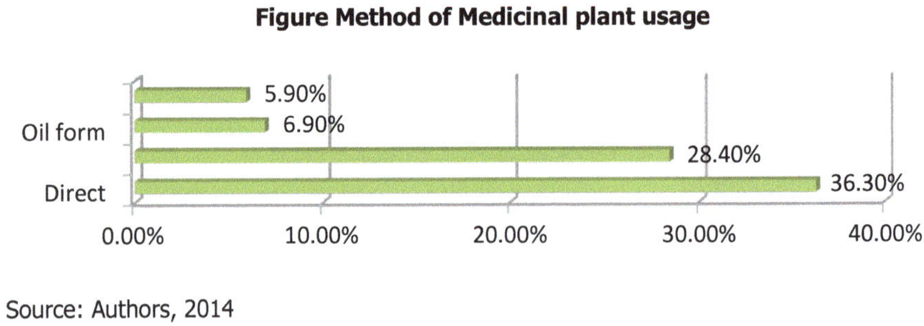

Fig. 4.6 Medicinal plant usage methods

4.3 Socio-Cultural Practices and Ecological Significance

India, a diverse country with deep rooted traditions, cultures and religions that prevail alongside, is one of the fastest growing economies in the world. The traditions and belief systems evolved over time continue to protect and ensure the well-being of the natural resource-dependent humans, animate and inanimate entities of this planet (Sitarama et al 2009). There exist several practices of offering prayers to plants/trees, ponds which help protect natural resources in this region. Documented records of the traditional practices provide us with insights into the practices prevalent over the years, changes in the current context and the need for their protection in the future. In this section, we specifically discuss the cultural practices of people belonging to Hindu religion since Hindus are dominant in the region as compared to Non-Hindus.

Worshipping Nature The traditional worship practices show a symbiotic relationship between human beings and nature. Indigenous communities all over the world even now live in harmony with nature, while conserving its valuable biodiversity (Sahu and Pankaj 2010). The sacred plants of India being worshipped throughout the nation can be attributed to their mythological significance. In India, various gods and goddesses are worshipped (Gadgil 2000) with plant flowers and leaves as part of pushpa puja and patra-puja. India has a deep-rooted tradition of nature worship, which provides a base for conservation of nature at the grassroots-level (Pankaj et al. 2013). The cultural/religious practices of offering prayers to plants/trees help protect and conserve diverse species. Another important aspect relates to the cultural tradition of processing and using these plants in the treatment of health related ailments at affordable costs. These practices are prominent under Jadigenahalli Gram Panchayat. The most common plants (generally found across the backyards/ front yards of houses) worshipped are Tulasi (*Ocimum Sanctum*), Alovera (*Aloe barbadensis*), Turmeric (*Curcuma longa)*, Brahmakamala (*Saussurea obvallata*) and others. These plants are grown for their medicinal properties, religious beliefs, and aesthetic values or as part of cultural traditions (see Table 4.4 and Plate 4.1).

Table 4.4 Religious significance and belief systems attached to plants/trees

Plant name	Religious significance in Hindu religion	Belief systems	Used for diseases
Tulasi	Represents Lord Krishna[a] or goddess Lakshmi[b]	A strong belief that prayers offered to Tulasi plant before leaving the house would bring in positive outcomes	Cold, cough
Thumbe	Signifies Lord Eshwara[c]	Used commonly for daily worship	Folk
Brahmakamala	Denotes Lord Brahma[d]	Belief that it represents god who resides in the plant	Spiritual and folk
Yakka flowers	Denotes Lord Ganesha[e]	Destroyer of all obstacles, hence, first prayers are offered to him during all functions – Religious and social.	Heat reducer/cooler
Turmeric	Denotes goddess Gowramma[f]	Worshipped during Deepavali (festival of lights)	Anti septic
Neem tree*	Denotes Mariyamma[g] and om shakthi[h]	During om shakthi festival, – Women and children sleep on the neem leaves – Make a bed of neem leaves and place their plates over it and have their food on the previous day of the ritual – Tie it around the temple entrance believing no evils will enter the premises – Hold it in both their hands and dance every evening during the festivals as a practice on the final day of the festival	Anti septic, anti viral used for infectious diseases
Peepal tree	Denotes Brahma, Vishnu and Eshwara	A belief that god resides in this tree, hence, worship it on specific days of the week.	Oxygenbars, good for married and pregnant women
Coconut tree	Denotes Kalpavruksha	All providing, every part of the coconut tree is usable, hence, denotes prosperity and all giving.	For treatment of all ailments treatment
Banana plant		Worshipped during Deepavali as it denotes positivity.	Digestive problems.
Aloevera plant		– Believed to sustain for long as a plant, hence, denotes longitivity of the household growing it. – After worship, Gel of the Aloevera is heated and placed on the temples of infants for good health. – Believed to reduce pollution	Skin treatment, women health

Source: Authors 2014

(continued)

Table 4.4 (continued)

[a]Lord Krishna is one of the important and popular Gods of the Hindu Religion
[b]Goddess Lakshmi is Vishnu's wife who underwent many incarnations alongside him as Sita with Rama or Rukmini with Krishna
[c]Lord Eshwara is another important God in Hindu Religion and worshiped as the principal Hindu deity, the destroyer of ego
[d]Lord Brahma the first of the Hindu Trinity, the God creator, he was the chief God during the period of Brahmanism, in entire India; only two temples are dedicated to him
[e]Lord Ganesha also known as Ganapati and Vinayaka is a widely worshipped deity in the Hindu pantheon. He is popularly worshipped as a remover of obstacles, though traditionally he also places obstacles in the path of those who need to be checked
[f]Goddess Gowramma signifies one of the forms of the main Goddess Lakshmi denoting Power and Strength
[g]Mariyamma is the South Indian Hindu goddess of rain. She is the main South Indian mother goddess. Goddess Mariamman and Goddess Kali are closely associated with each other
[h]Om Shakthi is Goddess of Strength

Source: Authors, 2014

Plate 4.1 Worshipping Tulsi plant

4.4 Sacred Trees, Temples and Ponds

In the ancient spiritual traditions, man was looked upon as part of nature, linked by inextricable spiritual and psychological bonds to the elements around him (Gupta 1971). Hinduism has always been an environmentally-sensitive philosophy. No religion, perhaps, lays as much emphasis on environmental ethics as Hinduism. Worship of trees and plants has been a documented part of religious practice in India since the hunting-gathering stage (Chandrakanth et al. 1990). Trees are decorated in India for a variety of reasons. Historically, sacred trees have been connected with rites of renewal, sexuality, fertility, conception, birth, initiation, death and rebirth. The tree is perceived as a form that houses the sacred, and thus is protected. Even the most rapacious Indian businessman would not dare cut down a sacred tree which is recognized through its adornment (Louise 2009).

4.4.1 *Ashwath Katte (Sacred Trees)*

Trees such as *Ficus religiosa, Acacia ferruginea, Aegle marmelos*, and *fius glomerata*, are commonly recognized by devout people in India as strictly religious trees. Out of which 'Ashwathkatte' is a combination of trees planted together for worship. Ashwathakatte consists of two Peepal trees and one Neem tree. We observed the area covered under each Ashwathkatte varying from 10×10 feet to 6×20 feet. There are 16 such sacred Ashwathakattes, conserving 48 trees.

A group of trees or a combination of specific varieties is planted in the temple premises, and these are popularly called as '*Arali marada katte*' locally (Kannada) or officially as '*Aswath katte*' named after a local deity, *Aswath narayana*. Similarly, there is a saying that "*mulatha brahma rupaya, madhyatha vishnurupini, agrathaha shivarupaya, viruksha rupayathe namaha*" meaning, the root being Brahma, the trunk Vishnu, and the leaves are Shiva (Chandershekar 2013). These trees are worshipped during all festivals and on a specific week day *Somavara* and (Monday) in a specific month (*Karthika masa*[9]). The belief system associated is these trees and the space surrounding them belong to god 'Devara jaga' and hence, trees cannot be cut nor can the space around them be occupied. This has helped conserve trees besides preventing encroachment. These trees are home to many bird species, squirrels, butterflies, honey bees (specifically in two villages – Govindapura and Jadigenahalli) (Plate 4.2).

Science behind peepal trees indicates that they release a fair amount of oxygen through the day, particularly during morning hours. Some plants such as Peepal tree can uptake CO_2 during night as well because of their ability to perform a type of photosynthesis called Crassulacean Acid Metabolism (CAM). However, it is not true that they release large amounts of oxygen during night (Rita et al. 2010). The ritual practised by several people in the villages relates to going around the trees in

Source: Authors, 2014

Plate 4.2 Worshipping sacred trees

[9]Kartika is a month in Hindu, Bengali and Nepali calendars, named after the Hindu god, Kartikeya. In the lunar Hindu calendar, Kartika begins with the new moon in November and is the eighth month of the year. The festival of Kartik Poornima (15th day Full Moon) falls in this month, celebrated as Dev Deepavali in Varanasi, also, the well known festival, (for the god of Sabarimalai, Ayyappan's garland festival.

Fig. 4.7 Sacred trees worship

circles during these specific days with the belief that it would bring good luck besides fulfilling one's wishes. Indirectly, it would help intake of more oxygen, instill discipline and bring about a positive attitude. It is also practised by expectant mothers and youngsters. It is interesting to see that across villages, more than 80% of the population worship these trees and 100% in the two villages (Jadigenahalli and KB Hosahalli). People are reported to have experienced calmness of the mind during worship. The main reason for tree worship among women is that, while unmarried women pray that they get married to a good husband, married women pray for a healthy child (Fig. 4.7).

4.4.2 *Ownership & Management of Sacred Trees*

There are sacred trees protected by upper caste groups as well as lower caste groups. Similarly, there are separate temples for upper and lower caste groups. Trees present in these temples are protected by the respective caste groups (31%). These caste groups own the temples as well as the trees. Besides these trees, there are sacred trees planted across common areas, worshipped and protected by all caste groups. Rest of the trees i.e. 68% of them are managed by all in the villages without being owned by any caste group (Fig. 4.8).

Growing and worshipping plants/trees has been a cultural tradition (47%), while 12% is attributed to as part of an ancestral practise; 20% of the people worship with a belief that it would bring to them general welfare. Besides, there are several other

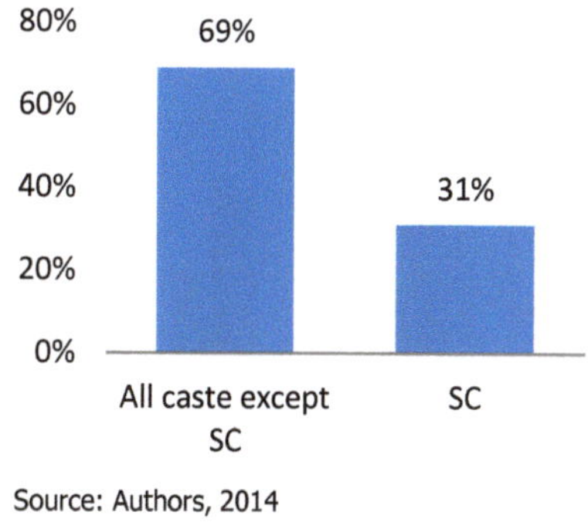

Fig. 4.8 Caste groups worshipping sacred trees

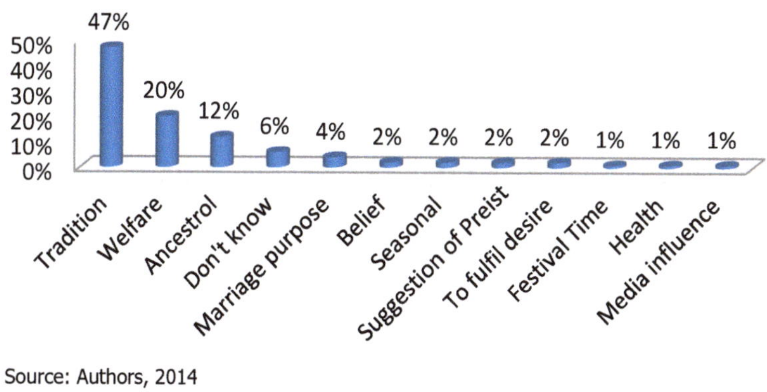

Fig. 4.9 Reasons underlying worshiping of sacred trees

reasons that influence people follow the practise to fulfill their wish list such as media influence, suggestions by priests, while there are a few (6%) who practise the ritual without knowing the significance. Irrespective of the reasons, one can ascertain that it is a good practice that helps ensure the health of a larger population and conservation of nature. The ecological importance of these sacred trees lies in the prevention of any usage; however, there are some reasons due to which 9% of this practice is violated (Fig. 4.9).

4.4.3 People Worshipping Trees

Plant/Tree worship is common and is practised by more than 70% of the people. Yakka plants are worshipped more as compared to Neem and Peepal trees. Unlike Neem and Peepal trees, Yakka plants are found near temples, road sides, back yards of houses and open spaces. However, it is interesting to note that all the three species are worshipped on a regular basis (Fig. 4.10).

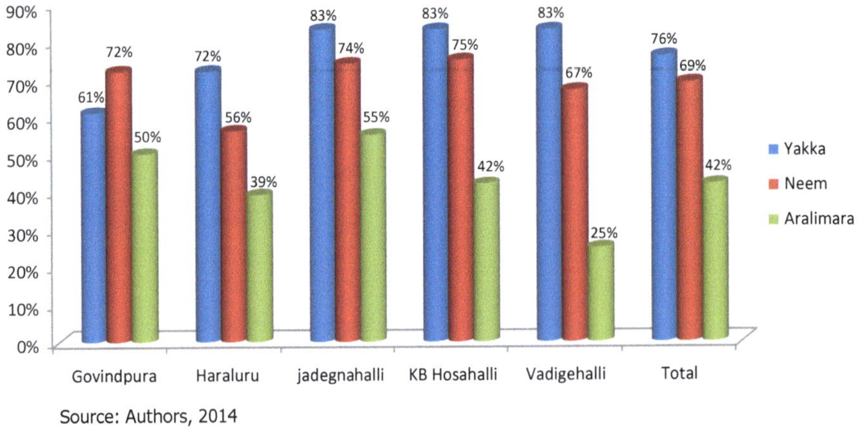

Fig. 4.10 People worshipping trees

4.5 Temples and Natural Resources Conservation

4.5.1 Village Temples

Many individual plants are considered sacred and worshipped as well as protected in different parts of the world. The continued survival of *Ginkgo biloba*, the living fossil in the plant kingdom is a typical example of the role of human faith and the conservatory role of religious places (Mohanty et al. 1997). In some villages, local deities (Grama Devatas) placed under some specific trees can be observed. It is a common practice in the rural and tribal localities of the state; these plants are also considered sacred and protected by the people.

Significance of village temples: Temple construction and conservation of natural resources are closely interlinked. There are subtle and obvious references made to practices followed on the lines of conservation of natural resources. Across the 19 temples documented, there is some form of association observed with nature worship (Table 4.5). It is interesting to note that they are also linked to mental health in that it is believed to provide mental calmness and peace. However, there are some temples to be comparatively linked more to natural resources conservation. For Example, Conservation of Land through tradition is seen in Haraluru wherein 100 acres of grazing land had been shared with Allemakanhalli in the past. However, this came under urbanization pressure, as land value increased. But local people fought to conserve this land in the name of "Bhupathiamma"[10] as they believed she lived in this place; this area, now under the control of the Government in the name of village treasure is being conserved as a sacred site. No development activities are allowed in this area. There are more than 35 varieties of trees in this area. Similarly, Goddess Guddadamma temple (Sapthmatheyaru

[10] "The mother of lord of Earth".

Table 4.5 Temples and belief systems

Village name	Sl no	Temples	Belief systems
Jadgenhalli	1	Karibeerana temple	Provide mental calmness
	2	Yellamma temple	Tradition and culture followed and preserved for future
	3	Eshwara temple	Religious beliefs transferred across generations
	4	Karagadevara temle	Conservation and protection of natural resources
	5	Gundudevamma	Ecologically important zones
	6	Sapthamathe temple	
Govindapura	7	Sapalamma temple	Once a year this god is worshipped and also cows are made to walk on fire. It is a cultural practice, but its main intention is to cure the foot wounds of cows.
	8	Kaveramma temple	All most all temples are believed to provide mental calmness
	9	Yellamma temple	Recreational purpose
	10	Chowdamma temple	Aids in continuing culture and tradition across generations
	11	Mariyamma temple	
	12	Saralamma temple	
Haralluru	13	Guddadamma temple	Sacred grove. Located on a hill top and is of high ecological significance – a green belt zone. Restricted area with no human intervention, except during the annual festival.
	14	Patalamma temple	Old temple worshipped by all
	15	Muneshwara temple	Conservation of land –green cover
	16	Sugalamma temple	Peepal tree called sugalamma, was worshipped regularly earlier. However, the tree fell as it was old and its space used for widening a road.
	17	Karagalamma temple	Located at the entry point of the village. Worshipped by all villagers, as they believe it is the goddess of the village
	18	Venugopalaswamy temple	Newly built temple for village recreation and cultural activities.
Vadgenhalli	19	Kalabiraveshwara temple	Historical value and cultural heritage
			Tourism spot

Source: Primary Survey 2013

meaning seven sister goddesses) located on a hillock has been allotted 400 acres of land around it with no scope for private development activities. This was initiated by the people in the context of development activities and encroachments taking place along with an increase in the demand for land as part of an urbanization process. In this context, the local people resented this development arguing that the place belonged to the temple of Goddess Guddadamma. Goddess Guddadamma is believed to be very powerful as seven gods 'Sapthadevatharu' are said to be residing in this area. In the name of this goddess and temple, local people fought to conserve this land and converted it into a treasury under the Government of Karnataka, India; so this area now under the purview of Mujarai (undertaken by the State Government) is being conserved with no development activities like the construction of buildings, factories which can be a hindrance to the conservation of ecology allowed. This, in turn, has resulted in ecological conservation, providing benefits in the form of fresh air, aesthetics and recreation and during cultural festivals, serving as a tourism spot.

However, the influence of urbanization and its consequences have a negative impact on the environment. The temple premises on the hillock are filled with garbage, mostly plastic paper cups, rendering the place unhygienic. The usual practice of using natural plates[11] and cups are being replaced by plastic plates and cups which do not decompose like plates/cups made of leaves. Similarly, the negligence of open wells located in the temples is also seen. For instance, in Jadgenhalli Mariamma temple, an open well (40 ft) is not well maintained; with garbage dumped, making its water unfit for drinking. In the past, the well served the drinking needs of the whole village. These are paradoxes, since, people are influenced by urban life styles, and they tend to use plastics liberally. Besides, the procedure of worship often creates waste, however, it is largely organic in nature and hence, can be used as compost material. It is interesting to note that efforts have already been made by village leaders towards maintaining the surroundings of the temple and also creating awareness in this direction.

Grama Devathe or Village Goddesses festival is popularly practised. Village festivals in the name of village deity are common in a majority of the villages in India. It is a practice which people follow diligently. It is a cultural folkway of improving social interactions and community wellbeing. Mariyamma is the local deity/goddess (Gramadevathe) of Jadigenahalli village. The village deity is believed to protect the village and well-being of the people. The goddess is worshipped by women during early mornings or evening hours. Special prayers are offered on Tuesdays and Fridays. On these days, women prepare lamps made of lemon and pray for her blessings. Similarly, special prayers are offered after Ugadi (Hindu's New Year), when they prepare lamps made of rice flour mixed with Jaggery paste (Plate 4.3).

People use Neem leaves during worship. Exposure to Neem helps protect people from diseases, as it is considered a disinfectant and is believed to prevent the spread of epidemics besides protecting animals from diseases. When children get measles,

Plate 4.3 Lamps made of Lemon (left), Lamps made of rice flour- jaggery – Ugadi festival (right)

[11] Natural Plates- (made of coconut/Plantain leaves).

Source: Authors 2014

Plate 4.4 Kala Byraveshwara temple in Vadgehalli

Neem paste is used to bathe them and is also believed to calm children when they act stubborn.

Kala Byraveshwara Temple (Plate 4.4), located on the outskirts of Jadigenhalli village, is believed to have been constructed during the Chola dynasty. Another belief is that it was built during the Pandavas times, more than 2000 years ago. Special prayers are offered to this deity on specific days in a month. On Ashtami day (the eighth day of Hindu lunar calendar), during Rahukala (A particular time of the day believed to be ruled by Rahu considered inauspicious), people light lamps made of pumpkin, (inner pulp is removed before being filled with sesamum oil). This ritual is believed to fulfill the devotee's wishes. Usually, it is the childless couples, unmarried girls and others as well who visit this temple. Besides, there are other rituals followed during the village fairs when Homa[12] is conducted. Young girls make lamps of rice flour and Jaggery and pray for their wellbeing. The temple is surrounded by mango, Peepal, Neem, Banyan, Wood apple (Aegle marmelos, Bale-local name) and Tamarind trees. This is another way of conserving tree species around the temple besides enjoying the natural/eco-services provided by trees.

4.5.2 *Sacred Ponds*

Karagabhavi is a sacred pond, worshipped as a cultural practice by all people across surrounding villages. A dedicated Goddess 'Karagadevi'[13] (Plate 4.5) is associated with this temple. The ritual, led by a family belonging to a certain caste group (Thigalaru), is a family tradition. Nearly 70% of the households belonging to this caste group are responsible for taking care of the sacred pond. The place is believed to be God's property and hence, not used for any other purpose. The temple

[12] Offerings to Fire God.

[13] Village goddess.

Source: Authors, 2014

Plate 4.5 Sacred pond as part of the cultural heritage

sanctorum which remains closed is opened once in 12 years with a grand festival 'Karaga' being celebrated. The festival is observed with utmost devotion by the village folk. The whole village will wear a festive look decked with flowers and mango leaves[14]*. Several informal rules are followed during this occasion viz. people are not supposed to walk around with slippers on for a period of 1 month; no vehicles are allowed within the village premises; no one is supposed to use umbrellas[15]; expectant mothers and young girls during menstruation are to stay in neighboring villages; people belonging to Scheduled Caste and Scheduled Tribes are to stay on the outskirts of the village, symbolically as guards of the village and so on. Goddess is taken to this sacred pond on the prescribed day and it is believed that the goddess bathes in the sacred pond, swims and drinks its water.

Although this story sounds superstitious, it has a mythological and ecological significance as people are restricted to use water from this sacred pond dedicated exclusively to the deity. Currently, the ritual continues to be observed with no water in the pond' though. With a severe water depletion, the sacred pond has not been full since the last 15 years, indicating a significant ground water depletion. During the festival celebrated last year (2013), the pond was filled by external sources through tankers. The sacred pond is protected by constructing a fence around it to ensure that it is not encroached upon and hence, protected. On the other side, the ecological benefit can be seen in the case of one bore well, located at a distance of 0.5 kms from this well with a water yield at a depth of 300 feet, which is the only current source of drinking water supply for the whole village. This is viewed as outcome of water storage in this sacred pond in the earlier times, one way of conserving water resources.

---

[14]Tying green, fresh mango leaves to the doorway is considered auspicious representing prosperity.

[15]As the temple sanctorum is opened only during this time, it is believed that God is in the village. Hence, to show respect and humbleness to God who is almighty, they observe this ritual.

Box 4.1: Significance of Using Mango Leaves in Festivals, Rituals

Mango or Aam tree is one of the sacred tree symbols of Hinduism. Since the days of the Puranas the mango tree is personified with various gods, goddesses and spirits. According to research scholars, in Aam Vriksha, Lakshmi, Govardhan, Gandharva and fertility god reside.

1. *Tying a mango leave "Thoranam" at the main entrance*: Mango leaves are used for the control of evil spirits. Mango leaves in bunches are said to absorb the negative energy from anyone entering your home. It is just to eliminate the drushti – i.e. the negative effects, at a place where the celebrations are held – be it at home or any public place. So it is placed at the entrance. Mango leaves have antibacterial activity against gram positive bacteria (Plate 4.6).
2. *Using mango leaves on the Kalasam and putting a cocanut in the middle of them*: Purna Kumbha mainly contains 'water' – The veritable life principle. Thus it is a direct worship of Varuna – The god of rain. He is the harbinger of rain which ensures fertility on earth and which nourishes animals and human beings. Apart from water, Purna Kumbha is filled with twigs or leaves of five trees – Ashwatha (peepul), Vata (banyan), Amra (mango), Panasa (jackfruit) and Bakula (Elengi). The mouth of the pot is covered with a husked coconut, which is decorated from the sides with mango leaves. Mango leaves are placed to represent fertility god. All the items used in the Purna Kumbha signify life in its bloom and plenty. It also indicates that human beings are part of nature and when nature thrives human beings too flourish.
3. *Using mango leaves for pouring ghee into the Homa Kundam:* For placing ghee, Purasu leaves are required. If it is not available Palasha (jackfruit) leaves can be used. If that is also not available, mango leaves can be used.
4. *Mangifera indica (Mango) also used in dietary food and its composition:* The unripe, fully developed mangoes of pickling varieties contain citric, malic, oxalic, succinic and two unidentified acids. The ripe fruits constitute a rich source of vitamin A; some varieties contain fairly good amounts of vitamin C also. Thc bark is astringcnt; it is uscd in diphtheria and rheu-

Plate 4.6 Mango leaves tied to the entrance of house(right) and the temple entrance(left side)

(continued)

Box 4.1 (continued)

matism; it is believed to possess a tonic action on the mucous membrane. It is astringent, anthelmintic, useful in hemoptysis, hemorrhage, nasal catarrh, diarrhea, ulcers, diphtheria, rheumatism and for lumbrici.

5. *Significance in treating different ailments:* The leaves are given in the treatment of burns, scalds and diabetes. Mango from the leaves has been reported to possess antiinflammatory, diuretic, chloretic and cardiotonic activities and displays a high antibacterial activity against gram positive bacteria. It has been recommended as a drug in preventing dental plaques. Mangiferin shows antiviral effect against type I herpes simplex virus (HSV-I). Source: Sri Vidya rajgopalan, significance of mango trees.

4.6 Summary

Several views by cultural anthropologists have illuminated the significance of culture and its role in combating climate change as discussed in the earlier part of this chapter. The above sections have highlighted the role of culture, traditions and their significance in the protection of medicinal plants, health and environment. Besides, Indian religions have been the advocates of environmentalism. They taught commoners an intimate contact and a sense of belonging to nature. Today when the world is undergoing a serious ecological imbalance and environmental degradation, it is all the more important for us to understand traditions that attach deep values to environment and sustain them. Working to strengthen cultural values is a precondition for the materialization of the necessary public demands for an ambitious government action on climate change. Strengthening these values will also persuade a greater obligation on the part of individuals to reduce their own environmental impacts. This chapter has provided evidence of ways in which traditions have played an important role in revitalizing health and conservation of environment besides influencing a positive quality of life of people. A culture-based plan will provide prospects for new and broad based coalitions between individuals and organizations to work in tandem to activate and strengthen inherent values. A wide range of actors can establish a common cause in promoting inherent values, irrespective of their particular social or environmental agendas.

As Hoffman puts it, political economy creates apathy for change. Social processes that guide people's thinking cannot be discussed without considering the economic, political, and technological realities that are both the enactment of our values and a source of inertia to changing them. There is a huge physical infrastructure around fossil fuels and the lifestyle they create, which cannot be replaced without problems. Also, there are strong economic and political interests that are in jeopardy by the issue of climate change (many of them controlling the infrastructure as mentioned). Resulting of this, they have adopted strategies to confuse and polarize the debate in order to protect their interests. Efforts to change the cultural views on climate change must include changing the vast institutions and infrastructure of our economy and must be prepared to deal with resistance from those who benefit from them (Hoffman 2015). Naomi Oreskes and Erik Conway in their book, 'The collapse of Western Civilization, focus on culture's pitfalls with regard to climate change. A fictional historical account explains that the onset of dangerous climate change and collapse of western civilization were brought on by market fundamentalism and excessively strict science based standards for taking a preventative action. To keep this dystopian future from unfolding, culture and conscience, have to change. He highlights that, culture and conscience are both adaptable; the question is whether we are up to the challenge of changing them. Environmental anthropology, and cultural studies of climate variability, offer key directions for future research and advocacy. Research has uncovered how drought, floods and particularly long-term changes in weather patterns are culturally mediated through practices, institutions, and knowledge.

Vital work has been done by cultural and human ecologists studying everyday subsistence activities and indigenous knowledge systems in small scale societies and among mobile groups (Ellen 1982; Netting 1993), and show how environmental perturbation is negotiated (Richards 1986; Minnegal and Dwyer 2000). In addition, this type of negotiation reveal substantial differences between the assessments made by scientific experts and those held by local people who draw their livelihoods from those environments. These practices need not be looked at as blind beliefs. It is up to the scientists to verify through laboratory experiments and research, the effectiveness of traditional practices and the underlying rationale influencing a reduction in climate change uncertainties.

References

AAA Global Climate Change Task Force (2014) Shirley Fiske and Tony Oliver-Smith, 11th Feb 2014

DM RC report Soukya foundation Jadgenhalli Gram Panchayath

Dr. Ashok Kumar Panda, Medicinal Plants use and Primary health care in Sikkim, International Journal of Ayurvedic and Herbal Medicine, Vol 2(2), p 253–259

G. Abhik, G. Kamalesh, Tradition and conservation in Northeastern India: An ethical analysis. Eubios J. Asian Int. Bioeth. **12**, 15–18 (2002)

http://agitsolution.com/cbse/ebooks/11TH%20CLASS/Knowledge%20Trading%20Practises%20in%20India/Chapter%206.pdf

H.J. Andrew, *How Culture Shapes the Climate Change Debate* (Stanford University Press, Pala Alto, 2015)

J. Barnes, M. Dove, M. Lahsen, A. M. P. McElwee, R. McIntosh, F. Moore, J. O'Reilly, B. Orlove, R. Puri, H. Weiss, K. Yager, Contribution of Anthropology to the Study of Climate Change, Nature Climate Change, Perspective, Published Online 2013, DOI:10.1038, NCLIMATE1775 (2013)

N. Jeetendro Singh, B. Singh and A. Gupta, Environmental ethics in the culture of meeteis from north East India, Silchar, Assam University

K. Pandey, K. Naveen Sharma, Traditional medicinal flora of the District Ghazipur (Uttar Pradesh, India). Int. J. Ayu. Her. Med. **2**(2), 307–321

M. Landauer, W. Haider, U. Pröbstl-Haider (2014), The influence of culture on climate change adaptation strategies: Preferences of cross-country skiers in Austria and Finland, Institute of Landscape Development, recreation and conservation planning, University of Natural Resources and Life Sciences (BOKU), Peter Jordan Strasse 82, A-1190 Vienna, Austria. Journal article : Journal of Travel Research 2014 Vol.53 No.1 pp.96–110

M. Andrew, J.O. Reilly, Anthropologists argue field must play a vital role in climate change studies. Nature Climate Change (2013)

S. A. Singh, A study of the relationship between Ecosystem services and human well-being In the coastal villages of the kubulau District in vanua levu, fiji, School of Marine Studies, Faculty of Science, Technology and EnvironmentUniversity of the South Pacific

Y. Vijaya Kumar, P.C. Sekhar, B.S. Lakshmi, S. Hara Sreeramulu, Folk medicinal plants used in the treatment of Asthma in Polavaram Forest area. Int. J. Ayu. Her. Med. **2**(6), 947–953

Websites

http://en.wikipedia.org/wiki/Geriatrics. (Accessed 29 Jan 2014/1pm)

http://en.wikipedia.org/wiki/Karnataka/. (Accessed on 11 Feb 2013/4.03 pm

http://en.wikipedia.org/wiki/Pediatrics. (Accessed 29 Jan 2014/3.20pm)

http://en.wikipedia.org/wiki/Rare,_Endangered_and_Threatened_(RET)_plants_of_Kerala

http://envis.frlht.org/ver_search.php

http://in.answers.yahoo.com/question/index?qid=20070320104543AAT1VMU. (Accessed on 12 Feb 2014)

http://transformed.businesscatalyst.com/media/articles/stakeholder_analysis.html

http://www.britannica.com/EBchecked/topic/1911274/Siddha-medicine. (Accessed on 22 Oct 2013)

http://www.census2011.co.in/census/state/karnataka.html. (Accessed on 11 Feb 2013/4.03pm)

http://www.dnr.sc.gov/species/pdf/TrackedSpecies.pdf

http://www.ecovillage.org.in/farm-fresh/culture-and-ecology/

http://www.indushealthplus.com/karnataka-health-statistics/. (Accessed on 11 Feb 2013/4.03pm)

http://www.jaim.in/article.asp?issn=09759476;year=2011;volume=2;issue=4;spage=179;epage=186;aulast=Chaudhary

http://www.medicalnewstoday.com/releases/93858.php

http://www.thefreedictionary.com/herb+garden/acessed on 13.02.2014

http://www.uib.no/filearchive/filetopic_paper-ecological-perspective-mcleroy.pdf

http://www.who.int/mediacentre/factsheets/fs266/en/

https://www.google.co.in/?gfe_rd=cr&ei=u1T7Uq6aJ6PO8gfm1YHQCA#q=home+remedies+meaning/. (Accessed on 12 Feb 2013)

Report by Dilli Homeopathic Anusandhan Parishad: Origin and Growth of Homeopathy in India: http://www.homeoint.org/site/delhihomeo/historyindia.htm. (Accessed on 18 Oct 2013)

http://www.thesolutionsjournal.com/node/969. (Accessed on 22 Oct 2016)

https://news.psu.edu/story/157057/2011/07/01/research/culture-influences-peoples-response-climate-change. (Accessed on 20 Jan 2019)

https://www.humansandnature.org/culture-big-oil-and-baboons-on-culture-conscience-and-climate-denial. (Accessed on 20 Jan 2019)

https://journals.openedition.org/eps/2397. (Accessed on 18 Dec 2018)

http://journals.sagepub.com/doi/abs/10.1177/0047287513481276. (Accessed on 18 Dec 2018)

http://www.futureearth.org/news/anthropologists-see-climate-change-human-problem-not-natural-problem. (Accessed on 18 Dec 2018)

https://news.ucsc.edu/2013/06/anthro-climate-change.html. (Accessed on 20 Jan 2019)
http://www.simonbatterbury.net/pubs/anthropologyandgw.htm. (Accessed on 20 Jan 2019)
https://www.globalchange.gov/content/social-science-perspectives-climate-change-workshop. (Accessed on 20 Jan 2019)
www.worldsocialscience.org/. (Accessed on 20 Jan 2019)

Chapter 5
Summary

Abstract This summary captures the essence of all the previous chapters emphasizing on urbanization and its implications on peri-urban areas, more specifically, Jadigenahalli Gram Panchayat in Bengaluru Rural District in southern part of India. The chapter succinctly discusses the changes in land use, rural health governance on public-private-people partnership, and at large ascertains the significant role of culture in conservation of natural resources and maintaining a balance between development and conservation from a broader perspective of human-nature synergy.

Keywords Peri-urban · Culture · Holistic health · Jadigenahalli gram panchayat

Urbanization is a dynamic and diverse process with a substantial scope for imbalanced growth of metropolitan regions, depletion of natural resources and the resultant climate change impact in and around the peripheries of cities. In other words, urbanization represents the diffusion of the typical characteristics of urban centres to rural surroundings causing externalities. However, the process of peri-urbanization happens largely in transitional zones between rural areas and cities (Iaquinta and Drescher 2000). Peri-urban areas are characterized by not only geographical, but also social and institutional transition. Socially, peri-urban areas are dynamic in nature, wherein social forms are constantly created, modified and discarded (Iaquinta and Drescher 2000). They are understood to be areas of social compression or intensification where the density of social forms, types and meanings increases, fomenting conflict and resolution. On account of land use change (against the backdrop of urbanization process) and the diversity of economic interests that this engenders, social groups tend to be heterogeneous and in constant transition (Allen 2003). Small farmers, informal settlers, industrial entrepreneurs and urban middle class commuters may all co-exist in the same territory, though with differing and competing interests, practices and perceptions.

Much of the social dynamism and flux comes from the presence of migrants and new settlers. These social changes and transitions further have spin-off effects in terms of the range of economic activities that they foster. Peri-urban settlements often draw migrant labor that seeks employment in adjacent towns and cities, adding to the heterogeneity of the population. It is also common for relatives of peri-urban

S. Manasi, K. V. Raju, *Coping Mechanisms for Climate Change in Peri-Urban Areas*, https://doi.org/10.1007/978-3-030-18517-6_5

dwellers to migrate to the peri-urban settlements in search of better living conditions, amenities or jobs in adjoining towns. (Narain and Nischal 2007; Narain 2009). This, in turn, leads to an increased demand for rented accommodation in these settlements with renting out of accommodation emerging as a major new economic activity. Besides changing land use and its implications for the peri-urban ecosystems, the rural communities also get influenced by urban culture (life style choices, in particular), more so in the peri-urban areas.

Institutionally, peri-urban areas are in transition as rural governance bodies may become defunct, without being replaced by urban governance bodies. Peri-urban areas lie outside the legal jurisdiction of cities and sometimes even outside the legal jurisdictions of municipal boundaries (Shaw 2005). This can contribute to the challenge and complexity of addressing peri-urban as a policy space, as many subjects requiring attention may fall within the jurisdiction of neither urban nor rural governments. For instance, in a study on Hubli-Dharwad in South India, the idea of a sewage treatment plant being set up was dropped, as it was not clear – who – the urban or the rural government would pay for it (Brook et al. 2003). Conflicts may also arise between rural and urban governments over issues of land acquisition for urban expansion, as witnessed in Gurgaon (Narain 2009) or between rural and urban claimants over resources like land and water as witnessed both in Chennai and Gurgaon (Janakarajan 2009; Narain 2009).

These peri-urban areas have attracted several studies focusing on peri-urban interface, investigating urban forms, spatial patterns, implications etc., however, there are few studies that have emphasized the role of ecological ethic and cultural practices in the context of development versus conservation conundrum, in avoiding the negative externalities of urbanization. The current study, which has made an attempt in understanding the 'Change' and its influence on environment in the peri-urban contexts, is an interesting documentation in this respect. It has provided insights in to the complex issue of urbanization and the associated externalities and brought to the fore the challenges faced by the peri-urban areas related to various aspects of health, environment and climate. In view of this, the current book specifically aimed at (a) locating land use change reflecting the impact of urbanization on water resources; (b) intervention of a governance model (public-private partnership) through Ministry of AYUSH for introducing holistic health care; and (c) promoting local health care traditions through prevailing cultural practices that play an important role in ensuring health and protection of the environment. The study has been able to ascertain that culture plays a significant role in protecting environment and thereby combating climate change implications extensively. As discussed earlier, the report by American Anthropological Association (Fiske et al. 2014) the AAA Task Force on Global Climate Change, along with a Statement on Humanity and Climate Change, highlights that the global challenges of climate change as being rooted in social institutions and cultural practices, demonstrating that solutions and social adaptations consequently require knowledge and insights from social sciences and humanities. Our study finding reiterates the same. In the study, we have also been able to ascertain the role of culture in protecting the environment and the possibilities of using such linkages with the interventions of providing public health initiatives and awareness. Cognitive filters reflect cultural identity. People tend to develop world views that are dependable with the values held by others within the groups with which they self-identify. Dan and Braman (2005), at Yale University terms it as cultural cognition wherein people are influenced by group values and will generally endorse the position that most directly reinforces the connections they have with others in social groups. It does not mean that scientific conclusions are rejected in the process, but are weighted and valued differently depending on how one's friends, colleagues, trusted sources or respected leaders value and frame these issues. People are the product of surroundings (chosen and unchosen) and gravitate towards opinions that fit with those they identify with. Hence, positions on topical and controversial issues like climate change become an inextricable part of people's cultural identity. Thereby, based on our observations, we have distilled out a few issues and their relevance to policy interventions that could be taken up in respect of peri-urban areas.

Jadigenahalli gram panchayat is representative of a semi-arid tropical region, in southern part of India, which could broadly represent a comparable situation in similar regions. To begin with, one can witness a significant amount of change in agricultural practices and livelihoods due to the influence of urbanization, a resultant change in land use. The implications of land use changes have been captured using Geographic Information Systems. In turn, this has also had a prominent impact on the environment of the region with respect to natural resources, particularly water resources. Ground water resources have depleted to a critical level, forcing to farmers to invest more money for procuring water. Water markets are thriving in the region with farmers purchasing water from tankers for cultivating crops. The trend is quite alarming because water purchase from tankers was usually restricted to drinking water earlier. A demand by the farmers to divert treated sewage water from Bengaluru city is a matter under consideration by the government. A shift in cropping pattern is quite prominent, moving from paddy to grape cultivation and floriculture. Focus is more adopting efficient water use technologies like drip irrigation is quite common across farms. Ground water status and related concerns analyzed highlight the severity of the problem in depleting ground water levels, failed bore wells, high investments towards ground water extraction, farmers' challenges, changing crop patterns and livelihoods with a specific reference to water scarcity challenges.

Understanding the collaborative initiative of the government through Public-Private Partnership to promote alternate medicines as part of influencing life style changes is another interesting component that we have addressed. Healthcare is the right of every individual, but lack of quality infrastructure, dearth of qualified medical functionaries, and a poor access to basic medicines and medical facilities has thwarted its reach to 60% of the population in India. A bigger challenge of providing rural health care in India has been bogged down by high costs and infrastructure challenges. Traditional medicine is a synthesis of therapeutic experiences from generations of practising physicians of indigenous medical systems in India. To build upon this traditional knowledge strength and also to address the healthcare challenge, the Indian government established the Ministry of AYUSH (i.e. Ayurveda, Yoga, Unani, Siddha and Homeopathy – which are diverse approaches in traditional medicine to improving health). The AYUSH Ministry promote the concept of AYUSH Gram wherein villages are selected for the AYUSH interventions of healthcare by setting up facilities including hospitals under the scheme of 'AYUSH and Public Health'. The Government of Karnataka with the support of Department of AYUSH, the Department of Health and Family welfare of the the Government of Karnataka and SOUKYA foundation, a charitable trust based in Bangalore, planned to establish the *Rural Health Care Centre* in Jadigenahalli, Hoskote Taluk, and Bangalore Rural district. This centre was the one of its kind in *Rural India* with integration of different systems of Medicine like Ayurveda, Yoga, Unani, Siddha and Homeopathy. The Soukya approach had turned out to be a positive development for economically poor communities in accessing health care facilities. In sum, the provision of quality traditional medicines based healthcare seems a good option to improving access to health care in rural India and other developing countries. Public-private partnerships can be an effective vehicle to promote the same. However, their financial sustainability post the period of partnership has to be reflected upon, and even decided, before implementation.

Urbanization has also made its influence on the sanitation aspects that could be attributed to life style changes in rural areas, thus impacting health aspects. Sanitation in the peri-urban contexts has posed several newer challenges like city waste being dumped in the vicinity of these villages. The challenges of managing newer forms of waste – plastics and its disposal – is becoming increasingly difficult. The unorganized way of managing waste has for sure affected the aesthetics of the village surroundings as well. Generation of garbage has increased with more plastic usage in the region. People resort to burning of plastic waste, causing air pollution. There is no regular collection mechanism to dispose of garbage, with several vacant sites being used as dumping yards which, in turn affects the health of those residing nearby besides polluting the environment affecting the aesthetics. Thus, dumping of waste from urban areas in several locations of the villages is another cause for concern. Resolving conflicts related to garbage disposal in the vicinity has been an issue for concern for the GP members. However, a positive change due to urban influence can be attributed to an increase in number of in-house toilets. Sanitation promotion campaigns by the Gram Panchayats and schemes by the government are being well utilized by the residents for construction of toilets. This also has aided in the village being recognized as a clean village besides being honoured with an award. Increased access to education with more youth studying in colleges has helped them cast off the practice of open defecation. Younger generations have insisted on construction of toilets at the household level and are using the toilets, thus, there are no open defecation prone locations. Open defecation is not acceptable by the community members and role of education and awareness can be seen prominently amongst the people.

Another notable dimension we have addressed is the role of culture in protecting the environment. Since Indian religions have been the advocates of environmentalism, the lessons have been taught to commoners in the form of traditions. With the current challenges of ecological imbalance and environmental degradation in the fore, it was interesting to document the cultural practices and understand the traditions with ancestral knowledge still being passed on by word of mouth through generations. Interestingly, it is observed that rural environment has been conserved in the name of culture and vice versa people depend on traditional knowledge and ecology of the region for maintaining their health status like worshipping of nature in the form of sacred trees, sacred ponds, rituals, village fests etc. Temples have played a major role in the conservation of local natural resources, like forests across large areas of land. Ancestral knowledge has helped transfer knowledge and message of conservation of natural resources and usage of it in daily life, as reflected in the case of medicinal plants and their usage. Medicinal plants have been cultivated and preserved in the back yard of households in villages and used for most of the health ailments. Through the practices of culture, there is conservation of natural resources like forest area, medicinal plants, trees, animals, landscape beauty, water bodies and mainly a sense of ecological ethic among the local people for conserving nature for present and future.

There are several aspects to indicate change; however, it is not possible to attribute the changes to climate change in particular. However, changes in various aspects of environment have been documented in detail highlighting the kind of impacts and influence that the peri-urban villages have been encountering. Changes in various aspects including land use, shift in cropping pattern, ground water depletion influencing livelihood patterns are prominently visible. In brief, we are able to indicate the on-going changes which could lead to further degradation and hence, needs attention. This research has also quantified the allocated agricultural land and shows that lands in the study area have declined to a considerable extent with severe implications for water resources. We feel, that urban and regional planners in developing countries like India should incorporate the use of real time remote sensed data and geospatial technology in monitoring urban expansion, particularly peri-urban areas, which currently is neglected. Besides, comparative studies of other peri-urban towns should be undertaken to develop scenarios and accordingly take policy initiatives.

Promoting governance through public-private partnership to be up scaled to other peri-urban regions would be a viable option. The implications of this intervention are significant in terms of improving the quality of life of the people, reducing medical expenditure and increasing in economic productivity. Identifying, documenting, protecting and promoting of the cultural practices prevalent in these local areas would aid in the protection and conservation of biodiversity, traditional knowledge and improvement of mental and physical health of the people. The cultural practices are more prevalent in informal ways could be streamlined to incorporate improved practices since newer ways of celebrations are in practice. Educating the people to ensure cleanliness and hygiene even after the celebrations would be useful. This is with a particular reference to waste management. There is a large scope for improvement in ground water resources management. Rejuvenating tanks would go a long way in improving ground water recharging and preventing encroachments. Rainwater harvesting through construction of farm ponds would also help improve ground water rejuvenation. Changes in Agricultural practices attributed mainly to scarcity of water can be improved/ overcome through usage of treated sewage water from Bangalore city by the farming community. Farmers have already taken to drip irrigation methods and are cultivating crops that are more suitable to the given water-scarce conditions; this may be strengthened by conducting more awareness training programs. Sanitation is a matter for concern and there is large scope for improvement in managing solid waste. Gram Panchayats can play a much larger role in solid waste management. Gram panchayats are to be supported by training programmes in solid waste management and it would be more useful to bring in more participatory involvement in this process. There is scope for making the areas completely free of open defecation practices, which is achievable.

In brief, we have been able to ascertain in our study the critical role of ecological ethic and culture not only in the conservation of natural resources, but also maintaining a balance between development and conservation, as also the importance of reviving and promoting indigenous cultural and healing practices from a broader perspective of human-nature synergy.

References

Allen, Environmental planning and management of the Peri-Urban interface perspectives on an emerging field. Environ. Urban. **15**, 135–148 (2003)

S.J. Fiske, S.A. Crate, C.L. Crumley, K. Galvin, H. Lazrus, L. Lucero A. Oliver-Smith, B. Orlove, S. Strauss, R. Wilk, Changing the Atmosphere. Anthropology and Climate Change. Final report of the AAA Global Climate Change Task Force, 137pp. December 2014. (American Anthropological Association, Arlington, 2014)

D.L. Iaquinta, L. Drescher, Defining peri-urban: understanding rural-urban linkages and their connection to institutional contexts. Tenth World Congress of the International Rural Sociology Association, (2000)

V. Narain, S. Nischal, The Periurban interface in Shahpur Khurd and Karnera, India. Environ. Urban. **19**, 261–273 (2007)

V. Narain, Growing city, shrinking hinterland: land acquisition, transition and conflict in Peri-Urban Gurgaon, India. Environ. Urban. **27**(2), 501–512 (2009)

A. Shaw, Peri-Urban inferface of Indian Cities: Growth, Governance and Local Initiatives. Econ. Political Wkly **40**, 129–136 (2005)

R. Brook, S. Purushothoman, C. Hunsul (eds.), Changing Frontiers: The Peri-urban Interface Hubli-Dharwad, India, Books for Change, Bangalore (2003).

S. Janakarajan, Urbanization and peri-urbanization: Aggressive competition and unresolved conflicts, the case of Chennai city in India. South Asian Water Studies **1**(1), 51–76 (2009)

K.M. Dan, D. Braman (2005), Cultural Cognition and Public Policy, Yale Law School, Public Law Working Paper No. 87

Index

A
Accredited Social Health Activist (ASHA), 97
Advance stock procurement, 116
Affluent residential areas, 11
Agriculture Man Ecology (AME), 96
Agroforestry, 70, 73
Area Planning Zones (APZ's), 15
ASHA Workers, 111
Ashwath Katte (Sacred Trees), 141, 142
Awareness creation, 99, 102
Ayurveda and homeopathy systems, 108
Ayurveda doctors, 115
Ayush Pushti biscuits, 105

B
Bangalore Metropolitan Region (BMR), 15
Bangalore Metropolitan Region Development Authority (BMRDA), 14

C
Corporate houses, 114
Cropping pattern, 68
Cultural schism, 125
Culture
  adaptation strategy planning, 125
  climate change impacts and adaptation, 123
  common health problems, 129
  cultural traditions/practices, 126
  Indian nationalism, 126
  local field work facilitates, 125
  medicinal plants, varieties and usages, 129, 130, 132–134
  natural and social sciences, 123
  organic cosmology, indigenous societies, 125
  pre-historic and scientific, traditional systems, 126
  public perception, climate change, 124
  religious preaching's, traditions and customs, 127
  traditional practices, 127
  vulnerable populations, 124
  worshipping nature, 138–140

D
Domestic and drinking water sources, 26
Drip irrigation, 66
Dr. Mathai's Rural Health Centre (DMRC) Soukya Foundation, 98, 99

E
Eco-Health' approach, 94

F
Farmers' perceptions, 57, 59, 63, 65
Financially sustainable options, 113

G
Gravity Recovery and Climate Experiment (GRACE) satellites, 45
Groundwater overuse
  capital losses, 44
  climatic conditions and man-made pressures, 44
  depth variations, bore wells, 57

S. Manasi, K. V. Raju, *Coping Mechanisms for Climate Change in Peri-Urban Areas*, https://doi.org/10.1007/978-3-030-18517-6

Groundwater overuse (*cont.*)
defunct bore well, 56
dircct and indirect socio economic driving forces, 43
functional bore well, 54, 55
GRACE satellites, 45
groundwater withdrawal structures, 51, 57
household and village level observations, 46
irrigated landholdings, 47
irrigation progress, 45
old open well near Kolathur, 53
old water storage structure (Kalyani) near Jadigenahalli, 54
open wells near Vadigehalli, 53
resource-livelihood systems, 46
self-regulating, 45
technological externalities, 45
unregulated exploitation of, 44
unused bore well, 55

H
Holistic health care, 156
Holistic treatment, 116
Home remedies, 116
Home remedy programme, 110
Hosakote Planning Authority (HPA), 15

I
Informal settlements, 11

J
Jadigenahalli gram panchayat
agriculture and horticulture, 20
ailments, 107
ancestral knowledge, 159
annual rainfall, 23
arthritis, respiratory disorders and hyper acidity, 108
AYUSH Ministry, 158
composition of, 21
diabetes and hypertension, 108
DMRC, SOUKYA Foundation, 106
domestic and drinking water sources, 26
fever, body pain and allergies, 108
gender status, 25
ground water status, 157
industrial areas, 20
literacy rates, 25
location of, 22
population and households, 24
rainwater harvesting, 159
respondents, 106
sanitation, 158, 160
semi-arid tropical region, 157
socio-economic status, 24
study area and location of villages, 20
taluk in Bengaluru Rural District, 20
treatment and cure, 108
types of housing under, 21
villages surveyed under, 23
villages under jurisdiction, 21

L
Lady doctors, 115
Land cover and land use
agricultural gardens, 42
agricultural land, 42
bore wells distribution, 52, 55–57, 61, 62
Brick Kiln near Govindapura, 38
built-up category, 33
classification scheme, 30, 31
commercial plantation, 43
Eucalyptus plantation near Jadigenahalli, 39
Eucalyptus plantation, agriculture land near Govindapura, 41
forest land, 42
GP-% distribution, classes across villages, 35
Grape garden near Jadigenahalli, 40
input data/materials, 30
intensive human activities, 30
Jadigenahalli area in acres, 33
Kharif Crop (Ragi) near Vadigehalli, 40
multi-temporal images, 30
1973, 36
rocky area near Haralur, 41
rocky surfaces, 43
satellite data, 29
scrubland near Jadigenahalli, 42
2009, 37
2013, 37
2003, 36
2017, 38
under Jadigenahalli gram panchayat, 34
unoccupied residential layout near Kolathur, 39
water bodies, 43
well inventory, 32
Literacy rates, 25

M
Medical equipments, 115
Medicinal plant usage and practices, 128

Medicinal plants
awareness creation, 103
usage, 136, 137
varieties and usages, 129, 130, 136
Micro Small and Medium Enterprises (MSME) industries, 10
Mixed settlements, 12
Mulching mechanism, 67

N
National Bank for Agriculture and Rural Development (NABARD), 10
Natural resources stock, 73, 74

O
Organic farming, 117

P
People worshipping trees, 143
Peri-urbanization
accommodation, 156
affluent residential areas, 11
anthropogenic activities, 1
anthropogenic pressures, 4
approach, 5
'built environment' and 'lived environment, 8
climate change dimension, 3
cognitive filters, 156
coping mechanisms, 3
ecosystems, 8–11
extensive land use and ecological diversity, 2
forestry, 14
gap, low-cost and social housing estates, 11
geographical environments, 3
human and natural resources, 5
hydrological, ecological and geomorphologic transformation, 4
imbalanced development, 2
Indian urban areas, 7
informal settlements, 11
linked deliberations, 2
LULC patterns, 6
marginalised and disenfranchised groups, 5
mixed settlements, 12
national, regional and urban planning activities, 4
over food and nutrition, 13, 14
Panchayats, 7
as place, a process/a concept, 4
residential and recreational areas, 5
rural governance bodies, 156
satellite towns and primary cities, 6
semi arid regions, 3
social and institutional transition, 155
state periphery areas, 1
traditional urban planning, 2
urban planners, role of, 8 (*see also* Urbanization)
water crisis and insecurity, 12
zones of contestation, 7
Personnel's communication skills, 116
Physical fitness and hygiene promotion
portable water and nutritional supplements, 104, 105
sanitation practice, 104
Yoga, 103
Population and households, 24

R
Respondents' coping mechanisms, 65, 66

S
Sacred ponds, 147, 148, 150
Sacred trees
Ashwath Katte, 141, 142
ownership and management, 142, 143
Sanitation process, 117, 119
Self-help groups, 114
Societal responses, 74–76
Staff strength, 115

T
Teachers, 112
Traditional Chinese medicine (TCM), 95

U
United Nationals Conference on Climate Change draft resolution, 94
Urbanization
development process, 14, 15, 18
research methodology, 18
survey design and instruments, 18, 19

V
Village leaders, 111
Village temples, 144, 146, 147

W

Water bodies, 43
Water protection and conservation, 117
Well-being, holistic health
  administrative, 112
  AME, 96
  Ayurvedic medicines, 96
  AYUSH project, 96
  curative, training promotive and rehabilitative services, 95
  and healing, 95
  indigenous/traditional healing systems, 94
  inter-linkages, 94
  people's cooperation, 112
  potential threats to public health, 93

Y

Youth organisations, 114

MIX
Papier aus verantwortungsvollen Quellen
Paper from responsible sources
FSC® C105338

If you have any concerns about our products,
you can contact us on
ProductSafety@springernature.com

In case Publisher is established outside the EU,
the EU authorized representative is:
Springer Nature Customer Service Center GmbH
Europaplatz 3, 69115 Heidelberg, Germany

Printed by Libri Plureos GmbH
in Hamburg, Germany